AGE OF SUICIDE

J.R. Kirsch

Publisher's Note: This is a work of fiction. Names, characters, places, and incidents are a product of the author's imagination. Locales and public names are sometimes used for atmospheric purposes. Any resemblance to actual people, living or dead, or to businesses, companies, events, institutions, or locales is completely coincidental.

Age of Suicide/ J.R. Kirsch. -- 1st ed.
ISBN 978-0-9974693-0-1

For my generation, and all who come after me.
For the love of life.
For the love of God.

HOLOCENE

Punching in a Dream

A chaotic mass of people gathered around the boarding area, tumultuous frenzy all around. The air was positively arctic, and clouds of gray frost could be seen emanating from every mouth with each exhalation. The murmurs of the crowd had intensified into a loud chorus of indiscernible shouts and screams, as two uniformed men fed thick braided ropes through winches affixed to the floor of the wooden deck. Cords tugged and pulleys creaked, and through the legs of the adults in the crowd, six-year-old Cooper spied a glowing white vessel descending ominously into the inky sea. It was one of the last boats.

The young child clung desperately to his father's hand. He had no recollection of where he was or how he had come to be here on this brilliantly lit rooftop, or was it a ship of some sort? There was no time to figure it out.

The crowd splintered into myriad zigzagging directions, each panic-stricken human as desperate for answers as he was. The general direction was *up*, as the deck was now sloping at an uncomfortable angle into the sea. Gravity was winning.

"Daddy, what's going on?" he screamed.

"We have to get to a boat," the father replied. A crown of curly brown hair sat atop the man's head, and his furrowed brow betrayed his deep concern that there were, in fact, no lifeboats remaining.

Cooper looked around frantically. "Where's mommy?" the child asked, wide-eyed at the thought of her absence.

"She went to get Allison. We have to stay here and wait."

The boy peered down the slanting deck. Flat piles of rope lay beside pairs of ghostly white hooks that jutted out from the sky, lines that had been used to lower the lifeboats from their davits earlier in the evacuation. A morass of people writhed at the bottom of the deck where the huge liner now met the water, splashing toward an overturned boat. The ship itself extended for an unknown length beyond the water's edge, which foamed white as the encroaching waves crashed upon the teak.

Cooper was taken in by the immensity of it all. He had no idea how he had gotten here or why everyone was screaming. He had no idea why his father, who customarily wore blue jeans and a T-shirt, was now clad in a rather stately brown topcoat, dark trousers and a shiny black top hat. Even his own appearance departed from what he was used to. What was this padded white vest that sat atop his six-year-old frame? It was clearly several sizes too large for him, more properly suited for a grown adult than a scampering child. His heart was beating furiously, like a carpenter pounding a nail into a board.

"Jeremy!" a familiar voice called from behind.

Cooper spun round to see a slender woman running towards him. Her elegant cream-colored dress rippled through the crisp winter air as she approached, and her long, powerful arms were carrying a child that clearly wasn't pleased about being awake at this hour.

"Robin!" cried the boy's father, releasing his son's hand to meet his wife. She lowered her unhappy daughter onto her feet, and outstretched her arms as her husband pulled her in to receive her. They shared a momentary embrace as their children watched beside them.

"What do we do?" she asked breathlessly.

"We need to find a boat, now!" yelled the father.

Cooper stared moon-eyed at his fear-stricken parents. He had never seen them like this before, so loud and desperate.

"I thought that was the last boat," Robin said, fear in her voice. "There aren't any more!"

"I couldn't hold it," Jeremy replied sullenly. "There might be more down front!"

Cooper could scarcely hear his parents' exchange; he was too busy observing the events unfolding around him. Ropes dangled from a pair of empty davits that once held a lifeboat. Framed between the cane-shaped poles was a middle-aged couple holding hands. It was a serene moment in an otherwise murderous scene, and they looked as two doves perched on a branch in the early morning breeze. They stared lovingly into one another's melancholy eyes, and shared a silent kiss that seemed to drag on for an eternity. To them, it probably was.

Then, almost without any forethought, they turned toward the open ocean and stepped effortlessly off the deck, disappearing behind the side of the ship. They were still holding hands.

Beyond the tumult of the passengers on deck, Cooper could make out a few pale shapes bobbing in the distance. He could make out little people in those white rafts. They were the feeble lifeboats rowing tirelessly away from the stricken liner.

He was beginning to understand: they were stuck on a hopelessly sinking ship, and there was no way out.

A pair of hands scooped him from the armpits; his father hoisted him onto his shoulders. Cooper now had the best view of anyone on deck, and he watched below as his mother took Allison into her arms. The parents began trotting toward the bow, now submerged under an unyielding tongue of saltwater.

Jeremy ran slightly ahead of his wife, dodging passengers as he galloped down the deck. He and Robin were the only two running against the current of people, trying to escape the mouth of the sea that was eager to eat them alive. The enormous funnels that once steamed merrily while the ship was at sail now stood dormant, the booms of errant furniture reverberating from the interior of the ship.

The pat pat patting of Jeremy's boots on the wooden deck kept Cooper's focus like a metronome, each step another beat in this terrifying symphony. He overheard the roar of the ship's propulsive machinery coming loose from the floor of the engine room, adding to the cacophony that amounted to the noisiest single moment of the child's life. And yet, over all of it, strands of genuine music snaked up to Cooper's ears. The velvety sound of violins could be heard weeping in the distance, and he looked and saw that his father was carrying him past a small ensemble of musicians.

"Nearer, My God, to Thee," he heard his mother say from behind.

Cooper threw his head up to gaze at the endless sky that overlooked the night. Each speck of light stared back at him, a million gleaming eyes watching out for his survival. Somehow he felt everything would be alright.

The next thing he knew, he was careening toward the floor of the deck. A sharp pain rang out from the back of his head, like that of a cinderblock bunting his skull. He lay stunned on the ground, holding his hands over the bruised cranium, now swelling into a tender bump. He let out a poignant scream that momentarily caught the attention of a dozen passersby before they quickly resumed their sprint for higher ground.

Jeremy scrambled to his feet. He had been knocked down by a stout man in overalls that wasn't paying attention where he was running; his full body check had sent Jeremy to the floor and Cooper flying, landing on his head a few feet from the side of the ship. The man ran off without so much as a word of apology. He was too busy saving his own life.

Cooper saw his father and mother rush toward him, but he was barely lucid enough to realize what had happened; he just knew that he was being carried in his father's arms toward the water, which was now only a dozen feet away. A number of men in officer's hats surrounded a boat with flimsy-looking canvas sides. It was clear that this was their destination.

Before they were close enough to board, the lines holding up the forward funnel of the steamer began snapping back. They shot loose like broken rubber bands into the sea, spraying up ice water as they hit the ocean. *Flit, flit.* One of the broken lines ensnared a passenger treading water nearby, whipping up blood and body parts. The massive iron cylinder groaned and heaved forward into the sea.

A giant tide swept up the slope of the deck, flooring Jeremy and Robin, and spilling their offspring into the hypothermic water. The wave brought with it dozens of helpless bodies floundering to stay afloat near the last lifeboat. The overturned hull of another canvas boat was pushed out to open sea, washing off the people that had been balancing on its backside. The helpless men, removed by the waves, swam back to their artificial island, clinging to its sides with numb, wet hands.

The seawater was like nothing Cooper had ever felt before. It was like being lacerated on all sides by some sinfully violent weapon. The cold left him paralyzed and knocked loose of breath. His wool coat weighed him down, and his eyes were stinging with salt. There were no thoughts. Only sheer terror and pain.

Jeremy swam toward his son's floating body. Cooper was crying loudly, joining the yelps of all others in the water around them. He couldn't spot his wife.

"Robin!" he shouted. He coughed seawater out of his trachea. The shock of the cold brought his mind to action, but his body felt limp and unresponsive. He and Cooper had been pushed toward the collapsible lifeboat, which now had people crawling up the sides, sloshing in water as they boarded. The rickety craft was in serious danger of sinking in its own right; nonetheless, Jeremy knew it was their only chance of safety, with the doomed luxury liner glowing fainter in the background as it angled into the darkness.

He grabbed his son and swam toward the boat with all his might. The men on board were operating brutishly, pushing off the ragged hoards trying to get in, the added weight threatening to swamp the boat.

Cooper looked around blankly, trying to ignore the throbbing pain of his bruised head and the bitter cold of his whole body. His discomfort was so unfathomable that he just wished it were over, that he was back in his bed in Florida, that he could forget this whole experience.

He was dimly aware that he was being raised out of the water now by his father's arms, and on the side of the dinghy he could make out a small brass nameplate: *R.M.S. Titanic.* He knew the death sentence that it meant.

His father slung him up over the gunwale and into a frosty puddle at the floor of the boat. A male passenger helped Cooper to a seat, where he plopped clumsily and silently. He swiveled his head to look over at the ocean from which he was now free. His father smiled up at him from the dark water, holding onto the side of the boat. Behind him was a stupendous sight, an iron giant of black and red, ants running to and fro on its topside. It reared its back into the air, pointing toward the heavens at an absurd angle. A yellow halo was enough to keep every detail in sharp relief, the magnificent electric lights obstinately shining on. And over the sound of the crashing dishware in the galleys, the huge boilers smashing into each other in engineering, and the groaning of the metal hull straining under the weight of the stern, it was the screams of humans that seemed most harrowing to Cooper. The extreme tragedy. The tremendous loss.

He couldn't shake the feeling that he had been through this before.

You knew this was coming, he thought to himself. *You knew all along.*

The lights on the great ship flickered once and went out. A final round of screams cut through the Atlantic air on a frozen April night. And then a crashing, gnawing sound, straight from the bowels of Hell.

"Cooper!"

The boy awoke in a fright.

"Cooper!" he could hear his mother cry in the distance. It was barely audible through his closed bedroom door, clearly originating from farther downstairs.

Immediately he noticed that nothing about this moment seemed right. A violent pat pat was hammering onto the roof above his room. A sensational hissing noise filled the air, a sound louder than a gale on a shoreside boardwalk. A current of howling wind swirled through the space, capturing loose papers and pens and sending them across the room. His square olive clock blew off the wall, crashing onto the bedside nightstand. The sound was followed by a booming thunderclap that resounded through the room.

The six-year-old had awoken from one catastrophe only to be cast headlong into another. Only this one was undeniably real.

"Cooper!" cried his mother again.

"Mommy!" he screamed in awestruck reply.

The air blew through his room, freezing on his face. He pulled heavy covers over his head. The call of his mom compelled the boy to move, but the whirlwind about him made him think otherwise. His cozy twin bed seemed a sanctuary, with his Mickey Mouse comforter nestling his small body like a cocoon. If there exists anywhere that a six-year-old can find complete safety and comfort, it's in his bed. It kept him safe from ghosts, from monsters in the closet, and maybe, he figured, it would keep him safe from a thunderstorm in his room.

But the sheer danger of his present situation demanded action, a realization that only occurred to him when he stopped shivering from under the covers of his roost. Before he could allow himself to cry, he closed his eyes tightly and took a deep breath. He then thrust his head up and out of his cotton kingdom to scan the room and ascertain what, exactly, was happening this time.

The nightlight by the bedframe cast enough of a glow to allow the child to see what had become of his room. The window that overlooked his plastic desk had been shattered open, letting out the thermic air that had protected him from the February chill outside. Through the razor edges of glass poked the arm of a tree, which now lay atop the flimsy desk, having cast aside the pile of books that had been resting there and spilling them onto the carpet. Blasts of wind and sheets of liquid were pouring through the breach, drenching Cooper's face with a spray of rain.

"Mommy!" he shouted again, louder this time.

He could make out no reply; the ambient noise was louder than a freeway.

Something stirred him to trade the safety of his trusted shelter in favor of movement, foraying into the hurricane that was blowing inside his bedroom. The bottom corner of his *Space Jam* poster flapped madly against the sky blue wall, waving him out to join the bedlam. Despite the noise and cold, curiosity took control over Cooper's actions, and he moved his legs step by step over to the gusting mouth that roared in his window. He approached the black hole slowly, using the broken tree branch to steady his quaking body.

He peered out. The sky was only barely visible against the darkness of early dawn. Faint pointed shapes could be seen against the dark twilight, heaving back and forth with the powerful wind.

Suddenly a spark turned the sky a bright white for but an instant, before receding to black once again. It was followed by a deep bellow, crackling with the untold fury of nature.

The solitary light in Cooper's room cut out instantly. Darkness consumed his compromised palace, and despite all his courage, the emotion that he had kept balled up in knots within himself at last unfurled like a tablecloth, blanketing him with a mortal fear.

Upset by all clamor, he plugged his ears with his fingers, jamming them in so far that they got stuck. The sharp pain was a welcome distraction from the weather.

The door behind him burst open.

"Cooper?"

His flustered mother sprinted into the room, her head locked in the direction of her son's bed. She couldn't make out much, but it seemed evident that he wasn't where she expected him to be. She ran her hands over the piled covers to feel if there may be a small person hiding underneath.

Cooper didn't notice her presence as he peered into the wrathful night. The air raged about him, but his fingers muffled all sound trying to enter his ears. He continued to stare out the hole in the wall, watching as blue flashes spread across the sky in rapid succession. The trees looked just as afraid as he was.

And then, just entering his field of view, a new shape: not a tree, not any manmade object he was familiar with, but a furious black coil that stretched limitless into the sky. The curving tendril snaked up into the clouds, bending itself left and right before thickening into a solid column of spinning debris. Large objects were being carried up by its immense power and flung off to either side, discarded with the same rabid intensity of a child tearing the paper off a Christmas gift.

Cooper had seen tornadoes before, either in movies or on the evening news, usually under the headline of "Extreme Weather" that so often seemed to lead the broadcast. But this was the first time he saw one in person, finding it every bit as scary as he expected. He let out a piercing scream.

Robin spun round in the unlit room. She spied her child kneeling by the branch that lay in the room like a severed leg, as if it had kicked its way through the wall.

"Cooper!" She rushed toward him and grabbed him around the ribs, hauling him up to her breast. She split for the door.

The boy wailed as loud as before. He removed the fingers from his ears and flailed about in his mother's arms, trying to turn himself around so he could wrap his arms about her neck. But her unrelenting grip wouldn't budge, and the best he could do was watch as she carried him through the rocking house. She ran across the hall and down the pinewood stairs, traversing them expertly from muscle memory, in defiance of her useless vision.

"Jeremy!" she shouted over the din.

"I'm in the living room!" her husband replied from below.

Robin pivoted around the railing and nearly faltered on the rug at the bottom, but continued her frenzied run through the cavernous foyer and into the living room. A solitary flashlight beam cut through the darkness, slicing up from the floor to her face, growing larger and nearer as Jeremy ran towards her. She met him in the middle.

"Is Allison safe?" she asked over the sound of the pounding hail.

"No, I couldn't find her! She isn't in her room!" came his harrowing reply.

Her eyes widened. She lowered Cooper to his feet, nearly dropping him, and snatched the flashlight from her husband's hand. She darted back toward the stairs.

"Get to the basement!" she called.

Jeremy acted immediately. He scooped Cooper into his arms and made for the cellar door with all his speed. Without the benefit of electric light, he was forced to envision the layout of his furniture, and estimate his position relative to it. After several near-misses, he veered too far to the right on his way through the kitchen doorframe, smacking Cooper's ankles against the hinge.

"Oww!" cried the boy, his head lurching up. He looked about in the darkness; all he could see was the dim sky through a wide, yet unbroken window. The enormous funnel in the sky still twisted upward, and it had grown tremendously. In his last glimpse before they left the room, he noticed the trees in front of the spinning demon arch backwards violently, before snapping at their bases and being sucked in.

Jeremy didn't register the events happening about him. He continued his relentless dash for safety. Cries for "Allison!" could be heard in the distance.

Robin had reached the top of the stairs and was back in the long hallway. The roar outside was growing louder. She aimed her light down the corridor, not knowing which room to start the search.

"Allison!" she shouted again.

"I'm here!" whimpered a reply. It was impossible to figure out where she was taking refuge.

"Where, honey?" Robin ran to each door, scanning the rooms with her beam. Not in the bathroom. Not in the study. Where was her child?

In fact, little Allison was hiding in the spacious closet in her parents' bedroom, buried under a makeshift fort of shirts, pants and jackets. This was a habit that she was thought to have abandoned by the time she turned three, resurrected for the maelstrom that had overtaken her house.

Robin continued searching. "Allison! Come here now!"

Jeremy and Cooper reached their two car garage from which he could access the basement. The blue SUV that carried the tech engineer home to his lakeside house each night now sat dormant in its chamber. Although his wife and daughter were still in harm's way, he knew he had to hunker down in the lifeboat of his basement, the entrance of which was now only a few strides away.

Jeremy hustled over to the ribbed aluminum door that covered the steps to the cellar. He dropped Cooper to his feet.

The pain in the boy's ankle caused his leg to instantly buckle under his own weight, and he dropped to the cement in a sad display, landing beside his mother's sleek red sedan. From the floor, he watched his father crouch to the door, fumbling desperately with the stubborn latch. The raging storm raised its voice to a clamor.

Then a loud crashing noise — Cooper looked between the wheels in time to see the wall of the garage crumple forward. A large brown tree trunk had asserted its way through the side of the house, rolling directly toward Cooper. It plunged right through the wall and blasted into the side of the SUV, embedding itself there. The car skidded into the sedan that sat beside Cooper in the next berth, and the two vehicles together absorbed the kinetic energy of the rogue palm. Crumbs of glass peppered the floor.

The tornado was upon the home. The tree was its knock on the door; it had now invited itself in. The funnel cloud entered the foyer obliquely, shredding the staircase as it tore through. A swath of drywall folded to the floor, raining insulation and electrical wiring. The uproar of splintering wood and flying glass deafened the house.

Robin heard the noise; she dived onto the floor. Allison heard her mother scream.

The twister carved through the hallway and moved toward the garage. The plates in the kitchen became Frisbees, the coffee table a hurtling missile.

Jeremy at last managed to unlatch the door. He picked up his son and marched swiftly down the stairs.

"Stay down here, Coop," the father commanded.

He turned back for the stairs. The aluminum door slammed shut in the wind, blocking off all light in the damp basement. Debris rained down onto the ceiling of their bunker. With the rest of his family amidst the ferocity, Jeremy made the heartbreaking decision to stay. Better to save two lives than just one.

Cooper was sobbing now. His father sat down next to him and brought his son to his side. Cooper buried his face in his father's chest as he cried.

"It's gonna be okay, Coop," Jeremy reassured. He didn't believe his words either.

"Are you just saying that?" the boy asked.

They sat there in the dark, two humans huddled together like gerbils, a tube of destruction passing overhead.

A minute passed.

Then several.

The horrendous noise dissipated quickly, but the terror remained. The thunder continued to clap, and the rain and hail fell with the same ferocity as before, so father and son remained in their bunker. But Jeremy knew the other half of his family was out there, and he had to go find them.

He gently removed Cooper's head from his torso and gave it a kiss. He shuffled about in the blackness, reaching around to get a sense of geography. A pole there, a table corner there, chair, bookshelf, binders, a lamp. He reached underneath the desk and searched through the drawers; he opened the lowest one. A collection of books, but nothing more.

He reached into the middle drawer. A pack of what felt like batteries. He reached further back, until he touched something round, smooth plastic on his fingertips. He grabbed it hoping for a flashlight, but it was nothing more than a tube of pencils. He grunted in frustration.

Top drawer. He yanked it open. More books, papers, a clipboard. But what was this? A slender box. He fiddled open the top. A cold metal cylinder lay inside, wrapped in a thin plastic bag. He pulled it out from the drawer and ripped open the casing. It was a mini flashlight.

Jeremy turned the knob at the bottom. Nothing happened. "Needs batteries," he figured. He untwisted the device and stuck his finger down the stem, confirming his suspicion. He reached back into the second drawer and fished out the battery pack. Whatever they were, he hoped they would work. He tore open the package, and he felt four dense rods fall into his lap. He grabbed one and dropped it into the flashlight; it fit. He quickly put a second one in and screwed on the cap. He flipped on the knob.

A shaft of light shot forth. Magic.

"I found a flashlight," he announced dully.

Cooper remained shaking, sitting uncomfortably on his bum and hugging his legs by his shins, tenderly avoiding his pained ankle.

Jeremy looked around the room. The black metal shelves behind him had a first aid kit of considerable size. He grabbed it and put it on the desk, unbuckling its brass locks. Then he paused and looked over at his son.

"Are you hurt?" he asked.

Cooper shook his head angrily.

Jeremy was unsatisfied by the answer. He aimed the light at the child and looked him over. Wet, a bruise on his left ankle, but thankfully nothing serious.

He went to close the first aid kit, but then he remembered something. He flung it open and removed a package of bandages. There, inset into the hard plastic, was a small hand crank radio. He removed it from its holster and began to turn the handle.

The crackling sound of classic rock radio began to hiss from the tiny speaker. He brought it over to Cooper and joined him on the floor. The beating rain still had not subsided.

"Here we go, Coop," he said reassuringly.

The smooth guitars were interrupted by the static of shifting stations. Jeremy was tuning to no frequency in particular, hoping merely to catch a snippet of something related to the weather. When he found nothing, he settled for NPR. It was 5:45 in the morning.

"Listen Cooper," Jeremy spoke, "I need to go up."

Cooper said nothing.

"Are you gonna be okay down here?"

The boy just sat, back against the wall, clutching his legs.

"Are Mommy and Allie dead?" he asked.

The question hit Jeremy like a blow to the gut.

"I'll be back soon," he said. He turned for the stairs, taking the flashlight with him.

The rain tap danced atop the metal door, making an irritating pitter patter. Jeremy turned the handle; the door wouldn't budge. There was a bowl-shaped dent in it that hadn't been there when they entered. He tried again, this time throwing his body against the aluminum sheet. It popped open.

He stood up and looked around, aiming his beam in all directions. Debris covered the scene. He peered back into the hole in the ground where his son remained.

"Stay here," he commanded. He closed the door.

In the monsoon about him, Jeremy trudged through the gnarled aluminum and sprinkles of glass that lay scattered across the remains his property. Through the driving rain, he heard the screaming of little Allison, who was suffocating under a pile of rubble. He sprinted over to the sound.

His sympathetic nervous system was in overdrive, his paternal instincts leading the way. His pupils dilated to full size. His muscles were taut with strength as his blood vessels directed oxygen to support them. The metabolic energy stored in in his cells was quickly liberated and converted into useful work, his arms straining tirelessly to free his wounded child. He threw a heavy board aside, and lifted the door to what was once his bedroom closet. Casting aside his torn apparel, he rescued his little girl.

He turned around and scanned the scene. Not twenty feet away, he could see his beloved wife's body, pinned lifelessly under a tree trunk.

Cooper sat alone in the basement. His shivering stopped, and the tears that had been flowing silently from his eyes stopped coming forth. He relaxed his body, straightening his legs against the floor. He reached for the radio.

"...with stocks poised to continue their growth through the rest of the quarter," came the calming female voice speaking perfect General American.

Cooper's mind dazed off. He couldn't focus on anything.

"...the phenomenon known as 'El Niño' continues to disrupt weather patterns across the United States. The Upper Midwest and Great Lakes regions have experienced snowfall levels that are well below normal for this time of year, and more severe storms are expected in California. Scientists have warned that precipitation will increase along the Gulf Coast and Southeast regions as the jet stream banks lower than normal, conditions which will persist as long as the pattern continues..."

That was fifteen years ago.

Help I'm Alive

High in the stratosphere, 90,000 aircraft cruise through American airspace on the daily. Their cylindrical engines draw air through powerful turbines, roasting it to 1500 degrees Celsius by the combustion of petroleum. The superheated air is shot out the back into a jet, propelling the crafts forward through the wonder of physics. Air traveling over their wings generates lift, carrying their huge fuselages into the sky.

That 400,000 kilograms of machine and metal could soar so effortlessly through the heavens seems almost miraculous, supernatural. Yet it is just an example of the endless capacity of human ingenuity, the conquering of flight one of the many achievements Man can take credit for. Despite the limitations of our biology, humans have managed to traverse the air as proficiently as a hawk. And we've done it just by using our brains.

It is on one of these flights that young Cooper voyages today, peering through a porthole to revel in the sight of the endless countryside below. Disregarding a wall of deep gray clouds in the distance, the view is serenely divine. Hill after rolling hill, a land of greens and browns, broken occasionally by the gray and black scars of human settlements. From so high up, entire cities look like mere crumbs on a green carpet.

There are thousands of people down there, he reminds himself as one such city comes into view. Time for a quick snapshot.

He pulls out his phone and pushes a small knob on the side of the slate, illuminating its mesmerizing rectangular screen. Placing his thumb over the biometric scanner, the phone instantly recognizes him as its authorized user, flashing the words "Good afternoon, Cooper," before unlocking the device. The display fades to his personal home screen, unfurling a neatly ordered array of candy-colored icons. He taps his thumb to the one labeled "Camera."

Before he has the chance to take a picture, however, low clouds obscure the view of the metropolis below. Disappointed, he puts his phone back in his lap.

It's his first time on an airplane, and until this moment, it's been an uncomfortable experience. Ever since he left Miami, when he said goodbye to his sister and father, he's been on edge. His chronic anxiety commands his thoughts, steering them toward worst-case scenarios that could rear themselves at any moment.

Cooper has an awareness, better than most his age, that one day he will die. The moment will come when he'll stop breathing, his brain will shut down, his eyes will close for good. Every step he takes is one more footfall in the direction of that inevitable end. He is a little more than a feeble biological machine, with a predetermined shelf life and a product cycle of planned obsolescence. He knows that his heart is only designed to beat so many times.

And just how many beats is his human heart endowed? A billion. A billion heartbeats. All mammalian hearts are designed with this limitation in mind. No matter who he is or what he believes in, Cooper knows that his heart will beat a billion times at most. And thanks to his overactive anxiety, it feels like he's used most of them already.

But he can't afford to waste any more beats. He's already 21; time is moving. That's why he's tearing across the sky like Zeus to New York City — to begin the first chapter of his adult life.

It's not that he didn't like Florida; it's just that three years of undergrad at his small community college wasn't doing it for him. He has bigger ambitions, which he knows Columbia University will fulfill. His first class starts in two days.

Perhaps calling them his ambitions is something of a white lie of which he tries to convince himself. They are, more properly, his father's ambitions for him. He may have grown up middle class, but Cooper was always aware that his family's prosperity was in jeopardy from a number of disasters that could crop up at any moment, condemning them down a path of starvation.

His father has yet to earn a college degree; he merely possesses the knack and maturity to eke out a comfortable life. But America in the 21st century is different from the America his father inhabited as a kid — without a degree, Cooper is professionally worthless. Student loans aren't cheap, so unlike his father, who started with nothing, he gets to start with less than nothing.

Hence his transfer to Columbia for his senior year. And hence the enormous pressure to complete his studies. By graduating as a business major, Cooper will validate his father's life, and achieve the prosperity that he knows he's earned. After all, business rules the world. If he is to inherit the full opportunity his country has promised him, that his father has pushed him to strive for, then he must become a master of the game. Anything short of that would constitute failure.

He's already been dealt one such failure in life — that wicked storm, when nature savagely robbed him of his mother. Fifteen years. Fifteen long years ago.

He's since gone through all the awkwardness of puberty, emerging thin and of average height, his ears a bit too big, but his childhood blond aged to a tasteful brunette. His bald cheeks still dimple when he smiles, and his large brown eyes telegraph that he's retained most of his innocence. He may be old enough to drink, but he barely looks old enough to drive.

Through the porthole, Cooper sees that storm clouds are almost upon the plane, replete with bright flashes that light their way through the dark layers. He may be flying through the sky in an aluminum chariot, but the threatening weather still leaves him with a sense of unease. Anxiety begins to take hold again.

To suppress his fears, he unwinds a pair of white earbuds from his carryon. He affixes them to the jack at the top of his phone, then carefully inserts the tiny speakers into his ears. Unlocking his device, he presses the music app icon. A black splash fills the display before revealing a series of tabs, each with a different heading: artists, genres, songs. He picks the last option and thumbs across the titles, a lengthy playlist of music both legally and illegally obtained over the internet. He selects a track.

Luscious dream pop drifts into his ears, sending his thoughts meandering back to his departure.

He remembers his teen sister, sheathed in a bright yellow summer dress, hugging him goodbye one last time before he went through airport security.

"I'll be back before you know it," he told her.

"There's no rush," she teased.

Then his dad extended a hand, which Cooper accepted with all the joy of meeting a car salesman.

"Work hard up there," he advised his son.

The boy nodded weakly, waiting for something more.

"I'll see you at Christmas," his father added, and with that, they turned away, leaving the young man alone to submit to the pat-downs. As he watched them slip past a corner and out of sight, he felt a nagging sense in his gut that he was about to face something he wasn't ready for.

But still the excitement — the thrill of being flung off on his own into the big bad world. A young adult. An educated man. Once he has his degree, the world is his. He can do anything. Be anything. He is, after all, a white literate male in the United States of America; how much more of a head start does he need? With a hand like that, to be anything less than prosperous would be something of a waste.

The difference between living in a small campus town and going to school in New York City is, of course, New York City. A whole world to itself. The chance for complete independence. He knows that this spellbinding moment in time — emerging adulthood — is the most exciting period of life, where anything is possible and nothing is guaranteed. Starting a life in New York may be a risk, but he's young and he's middle class and he can afford to take chances.

He can feel himself racing toward something bigger, the next stage. Brand new trials from which he'll learn, reveling in every glowing moment. He will drink lavishly from the juiciest vibrance of his life. This will be a year to remember.

His phone buzzes with a notification; he glances down to his lap. It's a text from his new roommate.

Shree: Hope you have a safe flight!! ☺ 3:19pm

He unlocks the screen and enters the messaging app, tapping out a reply with swift thumbs. A blue word bubble appears next to his avatar on the right side of the conversation thread.

Cooper: So far so good! I should be landing around 6. 3:20pm

An icon shuffles to the bottom of the conversation, a pair of eyes indicating that Shree has read his latest message. An animated ellipsis bobs up and down: she's texting him back. Her words appear with a buzz at the bottom of the thread.

Shree: Yay! 🎉 3:20pm

Cooper smiles to himself. He clicks on Shree's on-screen avatar, which leads him to her social media profile. "Shree Dunlap" is written at the top in bold sans serif font, her default picture inset beside it. The featured photo was taken from the Brooklyn Bridge, and it shows his roommate standing atop it, arms spread wide, facing the Manhattan skyline in the background. It looks like she's wrapping the whole city in a hug.

He scrolls mindlessly down her page. Post after post records every stray thought the young woman saw fit to broadcast, from words of wisdom to general status reports.

"Moving sucksss s," reads one from yesterday.

Another is an image macro of white text against a cloudy sky overlooking a valley, crepuscular rays beaming down from above:

> Shree Dunlap: Through the greatness and the beauty of creatures one comes to know by analogy their Maker 4:24pm

These pithy fragments of thought seem to be as much for her own sake as for the sake of her 579 friends and followers tuned in to her digital musings. Her page was one of the criteria Cooper used to judge whether or not he'd be compatible with her before solidifying his plans to cohabitate. If her feed is any reflection of her character, Shree seems like someone he will get along with nicely.

He flicks his thumb down the screen like a whip, the text flying upwards to the very top of Shree's profile. He taps the search bar and begins to type. M-E-L-A—

He clicks on a name from the suggested options.

It takes him to Melanie Livingston, his other soon-to-be roommate. Compared to Cooper, she seems more than a little exuberant. All of her pictures look to have been snapped at rooftop parties, raves, and bars. Her social media presence is enormous: with over 1400 friends, she makes Cooper self-conscious about his own paltry 283 connections. He wonders how many of those people she actually knows.

Nevertheless, she looks like an interesting person, and living with her should yield a lot of lessons about how to have fun. With a little luck, this girl could be key to enjoying New York.

It's no wonder she's so popular. Scanning the profile pictures tiled down his screen, he can't help but notice how attractive she is. Long blonde hair, stately eyes — she looks like a rich white girl, alright. Her statuses are a bit more provocative:

Melanie Livingston: Time to turn upppp – at Pour House 11:07pm
Melanie Livingston: Sangria and muffins with bae 10:51am

And then the selfies; oh so many selfies. In each one, she's either cheesing into the camera with her boyfriend or hanging off some bro stereotype's shoulder in a nightclub.

Cooper chuckles to himself as he considers the implications. Hopefully someone this rowdy will turn out to be a respectful roommate.

The music drags on in his earphones. He's finally relaxed, his mind drifting off to sleep. As he slowly enters a dream, he doesn't notice the slight trembling that gradually vibrates through the plane. But then a sharp lurch.

Cooper is shaken awake from the movement. The shuddering continues, more adamant this time, and the anxiety returns.

The plane rocks in the thunderclouds it's flown into. Cooper throws his gaze out the window: pockets of flashing light harken to that frightful night fifteen years ago. As the turbulence continues, he can feel his stomach turning itself into half-hitches. The power of nature is coming to finish him off. It will knock the plane out of the sky, swatting him down like a gnat.

It's just turbulence, he assures himself. He bites his lip and closes his eyes.

The plane continues to bounce; he can feel nausea coming on hard. His heart is now pounding, like an angry dog trying to bark its way out of his rib cage. His breathing quickens audibly. Another sharp bump, and Cooper grabs his seatbelt in a panic, jamming it into the fastener and accidentally brushing the muffin top of the portly gentleman seated beside him.

Seeing this, the fat man turns his attention from the newspaper.

"Hey kid, you alright?" he asks.

Cooper opens his eyes, darting over to the burly man. "I think so," he says. "It's nothing serious, obviously."

"Of course not," the man assures. His wide face looks trustworthy enough. "Turbulence is nothing to worry about. Trust me. I'm a pilot.

Turbulence is expected."

The shaking subsides marginally, and Cooper feels himself regaining control of his thoughts. He slows his breathing, and he listens as his heart rate starts to fall. He looks over to the seated man.

"How long have you been a pilot?" he inquires, taking his mind off the experience.

"Over thirty years," the stranger answers. "Turbulence happens. Don't let your nerves get the best of you."

"I feel like every time I see plane turbulence in the movies, it's like the most horrifying thing ever." Cooper unlatches his seatbelt reluctantly. "It always scares the hell out of me."

"It's actually pretty common, especially nowadays," allays the pilot.

The boy's eyes widen. "What do you mean nowadays?" Cooper asks. In what kind of mad world is turbulence becoming commonplace?

The man folds his newspaper into his lap. "Turbulence is usually caused by bad weather, like that nasty thunderstorm out there. Storms that big are becoming more common, so if you're planning on flying back and forth between New York and Miami on the reg, you gotta be prepared for a bumpy ride." The pilot turns back to his newspaper, his eyes picking up where he left off.

Cooper gulps hard. He looks back out the window, the nausea returning at the sight. It's unlikely he'll ever be flying again.

. . .

A group of casually-dressed young adults are seated about a large round table, occupying all twelve of the rolling chairs arranged beside it. They sit facing a large screen on one end of the room, which provides the only illumination in the dark cellar.

A young man stands by the display, leading his audience through a slideshow. He gesticulates with his arms, navigating through each portion of the presentation in a series of movements that push and pull new images into view. He has everyone's complete attention.

"I've been following Coke with my camera for months now, capturing key moments like this here, picking his nose," he says, pointing to an unflattering photograph of a wrinkled, white-haired man with his index finger in his nostril. A chuckle rings out through the room.

"Treat him like a celebrity, right? That's what you told me to do, and I have," says the young photographer. "We've blogged about his values. We've called him out on his harebrained ideas. We've tracked his Coalition's stance on every issue. He is our celebrity of evil."

The crowd nods in silent agreement with the paparazzo.

"In the past few days, however, I've noticed something disturbing." The photographer motions his hand left again. An image slides onto the screen at his command. The picture features two small figures walking down a street. The presenter brings his hands together and pulls them apart slowly, directing the computer to zoom in on the people in the photo. One of them is the white-haired man. The other is a small, bespectacled gentleman with brown skin.

The men and women squint at the screen with uncertainty. None of them seem to recognize this shorter man, who appears to be chatting genially with the older gentleman known as Coke.

"And *he,*" the young journalist says, pointing to this mysterious figure, "is the reason we called you here today."

"Who *is* he?" asks a woman seated at the table.

"I can answer that," speaks an assertive voice from the back of the room. She moves to the front, nodding to the presenter that it's her time to take the stage. As she approaches the screen, her chocolate face captures the light, flashing over the dark gentle hair that sweeps down the sides of her cheeks and revealing her delicate features.

"Thank you, Indu," she says.

"No problem, Sister Nevaeh," the photographer replies, taking his seat.

The woman turns to her audience. "Through face recognition software, we've been able to identify him as Doctor Jorge Valdez." Nevaeh pulls her hand through the air, revealing a detailed summary of the man in something resembling a medical file. A portrait in the upper right corner of the screen is framed by lines of text, each sporting details about this otherwise unknown gentleman.

"He's a leading Mexican aerosol specialist, currently employed by a highly secretive research firm that receives funding from Coke. Beyond that, we know nothing."

A palpable tension has filled the room. Several heads turn toward one another, mouths open in frozen gasps. They can see the whites of each other's petrified eyes.

"Yes, I know," says Nevaeh, addressing their unspoken thoughts. "This could be a serious problem."

"So the rumors are true," says one of the delegates.

"You really think he's planning something?" asks another.

"At this point, there's no way to know," Nevaeh responds uncertainly.

"That's not *good* enough," interjects an alarming voice. All eyes turn to the man, whose hard face glistens uncomfortably. His black, downturned eyes peek up suddenly, scanning the faces in the room.

"If he's working on what we all think he is, we need to be certain. We have to investigate this." His words cut with the force of a dagger, bringing voice to everyone's concerns.

"Well, Santi, that's the other reason why we've called you here today," Nevaeh says, "because we have a plan." The captivated eyes shift back to her. "But it's not going to be cheap."

"Who *cares*?" asks Santi, raising his voice to a shout. "If this *is* a geo-weapon, the cost is irrelevant."

"I agree," says a woman across the table. Her straight red hair is drawn into a pony tail which protrudes from the back of the Philadelphia Eagles ballcap on her head. "We need to know what the Cokeheads are up to. What's your plan, Nevaeh?"

Nevaeh gestures to a new slide, depicting a satellite image of an inconspicuous warehouse-like building. "We've been scouring the darkweb for leads. So far we've been able to trace aerosol shipments to this location in eastern Mexico." She points to the gray structure on the screen. "Satellite mapping indicates a compound there that's been constructed within the past six months."

"Do you have floor plans?" asks Santi adamantly.

"Not yet, and we likely won't find any through our usual contacts," says Nevaeh. "We're going in blind on this one."

"That won't end well," warns the redhead.

"Believe me, Val, we know," Nevaeh laments. "But you're both right. We have to investigate this. We don't have a choice."

The room is silent. They know the gravity of the situation.

"Two agents will infiltrate the building and install a surveillance package on their mainframe," she continues, holding up a flash drive. "Once it's on their computers, we should have a better idea of what the Cokeheads are up to."

More nodding. One of the delegates volleys the inevitable question: "Who's going to go?"

Nevaeh eyes the pair of newcomers that sit with them at the table tonight. They haven't uttered a word yet, preferring to let the facts speak for themselves. They return her acknowledgement with a nod.

"Brooklyn Project will lead the charge on this one," she announces, motioning to the agents. "I've selected Tyrone and Iri, who have joined us tonight. They've had their training, and they know the stakes. They've pledged a full commitment to this mission."

The congregation studies the two young soldiers, both in their mid-20s. Each of them silently appraises their capability in carrying out the dangerous objective.

"If you trust them to get the job done," Santi declares, "then we're behind you all the way."

Nevaeh smiles. "Then we'll execute as soon as we're ready. But before we do that, we need a contingency."

"Always have a Plan B," Val agrees.

"If this op doesn't work, you know they'll be gunning for us." Nevaeh pauses for a moment to emphasize her point. "We need to be in complete solidarity here. I'm not putting BKP in jeopardy if you guys don't come to our defense when things turn ugly." She scans the faces in the room; they appear to understand.

"You want a mutual defense agreement?" asks Santi.

She nods. "An attack on one is an attack on all. The people in this room represent the heads of all partner decarb organizations. If this sparks a larger conflict, we need to be united. This could be it, folks."

Val nods aggressively. "Count Philly in. If Brooklyn Project comes under fire, you have our support."

Santi grimaces at the familiar faces at the table. "I think we all agree this is too important for you to carry the torch alone."

Nevaeh walks over to a desk beside the projector screen. From it she grabs a tablet, pulling up a text document onto its screen.

"I have here a charter that would bind us together for the stated goal of aggressive decarbonization, at any cost. The Clean Front, if you will. It outlines the stipulations of our alliance, and justifies the use of force to that end. I'd like each of you, as representatives of your respective organizations, to read it and sign it."

Santi scoffs cheerfully. "You don't have to convince me," he says. "Just let me have it so we can get this show on the road."

She passes him the tablet. He depresses his thumb into the fingerprint sensor for two seconds, providing his signature. He then passes it to his left for the other delegates to follow suit.

Nevaeh watches approvingly. "While that's going around, it's time for a congressional update. Yari, how are things on the political front?"

A man in a green polo raises his head, drawing attention his way. "Well, the new carbon bill has stalled in the House. I'm trying to convince some coal state Democrats to support it, but for now it doesn't look like they're gonna budge."

Santi snarls. "Oh, forget about the political front already. You know there's no progress being made there," he growls bitterly.

Yari swivels to face him. "If we don't get even a basic emissions reduction framework in place before the climate conference, how are we gonna convince the rest of the world to follow suit? We don't have a lot of time left."

Santi throws his hand in the direction of the screen, still depicting the satellite image of the Coke compound. "With all due respect, Representative Hansen, but if this aerosol business turns out as bad as it looks, we may have less time than we thought."

"As usual," Iri mutters under her breath, her first contribution to the meeting.

Yari turns toward the projection. "Can't argue with that," he mourns.

"Okay, enough," Nevaeh quells. "Yari, keep working hard, as always. As for the rest of us — we're in agreement?" She looks around at her compatriots.

Val removes her thumb from the tablet, and she hands it back to Nevaeh. All the delegates have signed. They nod at her grimly.

"Alright then. Let's save the world."

The Night Is Still Young

The airport flurries with activity, people moving in all directions. Most follow the traffic pattern by walking on the right-hand side of the corridor, although there are always a few exceptional individuals that fail to understand this rule. The fluorescent light soaking the scene from above is not a pleasant experience for Cooper, but he soldiers on as always.

He emerges onto the street with his black duffel in tow; a fleet of yellow taxis waits impatiently for money. Eager to get to his new apartment, he hails one and approaches it. The driver pulls a lever under his seat, opening the back trunk. Cooper throws his bag inside and climbs into a cab without an adult for the first time. He privately wonders if he's about to be kidnapped.

Cooper stares out the window at the sprawling metropolis around him. Interminable stretches of concrete and tarmac lead him through the neighborhoods, and there are people, oh so many people, everywhere, on every corner. Children run through water spraying forth from a loosened fire hydrant. Adults of all ages sit on stoops that protrude from every apartment, talking and enjoying the company of their neighbors. Ice cream trucks play happy music for the streets, interrupted by the occasional emergency siren. This is summer in New York. This is Cooper's new community, his new home, where he'll build a life. It's make or break, and break isn't an option.

The taxi drops him off at a corner right on his block in Bushwick. At nearly fifty dollars, the trip cut deeper into his modest funds than he anticipated. After a minute on these crazy New York streets, Cooper reaches his new abode, a tall rectangular apartment complex with straight walls of cement. Its dark gray sides are pockmarked with even rows of square windows set flush with the exterior facade. Each of the four steps to the front door sport the same utilitarian concrete as the rest of the building. Taken as a whole, it may be the boxiest sight Cooper has ever seen.

He climbs the stoop to the aluminum door, the wheels of his duffel smacking every successive step. He texts Shree his arrival, and pushes a polished silver button on the side of the building, labeled "apartment 4G" in penciled scribble. The bell rings urgently; after a few seconds, the door emits a low buzz, unlocking itself as Shree lets him in from upstairs. He pushes it open for the first time.

Inside is a large, industrial staircase, with black handlebars and navy blue rubber steps. It slinks upward seemingly forever, illuminated by narrow lengths of hanging lights over the landings. Cooper inhales deeply and trudges up the first bank of stairs, hauling the bag with him. Each clap from its wheels bounces off the sepia walls, echoing loudly through the atrium.

Three stories later, he reaches his destination floor, winded but alive. He rolls along the hallway over the worn hardwood floors, passing the anonymous doors of his new neighbors. Finally he reaches his own dwelling, 4G. He leans the duffel against the door, pausing for a moment to catch his breath. He hears footsteps through the door, so he raises his hand to knock.

Before his fist makes contact, the entrance swings open, sending his luggage tumbling to the floor. An astonished young woman stands in the narrow foyer, her solid black hair playing in loose curls about her shoulders. She looks down at the toppled bag with surprise.

"Oh, goodness, I'm sorry!" she hollers, reaching down to pick it up by the cloth handle.

"Oh, here, I got it," he offers, taking the leaden sack from her with a smile.

"Jesus, is everything you own in that bag?" she asks jokingly, noting its exceptional weight.

Cooper laughs mildly. "It sure feels like it," he says, exhaling loudly as he lifts the bag upright. Now balanced steadily on its wheels, Cooper lets go and holds out his hand for his roommate. "Anyway, I'm Cooper."

The girl in the doorway returns his handshake. "Shree," she says, a wide grin curling up her dark freckled cheeks. "Welcome to our apartment!"

The sincere joy in her voice makes Cooper instantly feel welcome, and he smiles into her warm round eyes. She turns inside and waves him in her direction. "Come on in," she gestures. "I'll show you around."

A thump from behind them; the bag has fallen again. Shree giggles, triggering Cooper to do the same. He reaches down for it.

Already this building is a far cry from the dorm rooms he's accustomed to. This ain't no amateur hour; this is an *apartment,* something *adults* live in. He's left the nest. He's on his own.

"So here's the kitchen, as you can see," she guides, motioning to the white tiled space set off to their right. The kitchen's open borders are defined by the structural pole opposite the wall corner, beyond which the tiles on the floor give way to bare concrete. The low ceiling above their heads appears to be nothing more than the exposed undersides of the floorboards of the upper level.

The lip of the ceiling rises up past the pole, forming the sides of the rooms above. The space beyond is a towering chamber with naked walls; a window lives on the far side, although no light seems to fall through it.

"This is the living room," she says, referring to the cavernous den. It is furnished by nothing more than two sofas facing each other, a cheap black table between them, and two large throw rugs by their feet. To the left is a rope ladder that dangles from the kitchen ceiling, providing access to the open upper deck. It looks like a tree house that was originally just a flat wooden floor, until three partitions of thin drywall were erected into crudely-formed spaces. Slender doors and a narrow hallway provide access to these small rooms. The weathered wood railing cages the whole space together, with the word "LOFT" etched into its round handle.

"And up here are our rooms," Shree says, stepping onto the rope ladder.

Cooper follows her up the sturdy rope. Its brown fibers bite into his palms like microscopic syringes, irritating his hands as he climbs. Shree offers him a hand for support as they right themselves on the narrow wooden platform.

"Thanks," Cooper breathes.

"You're welcome." She continues forward, passing one door and stopping just beyond the second. "And this is your room," she says.

Cooper grabs the brass handle and swings the door toward him, driving it hard into the wooden railing that quite obviously doesn't provide enough clearance.

"You have to be on this side," Shree points out from behind the door.

"Right," says Cooper, feeling like a dunce. He closes the door and joins her on the other side before opening it again. Inside is a space only marginally larger in area than a walk-in closet. A twin-sized mattress consumes half the floor, lying there without a frame. Its once white fabric has been worn into an off-putting beige, with a darker brown stain that spills over the side, possibly the product of vomit thrown up however many moons ago, if not just last night. Beside it is a black table of identical design to the one in the living room downstairs. An outlet is screwed crudely into the drywall, wired to God knows where. This is his new room.

"Alright, cool," he says after a moment. He turns back, having seen all there is to see. He steps back onto the creaky narrow hallway and closes the door, allowing them to pass.

A fire would kill us all in about ten seconds in a place like this, he thinks. But maybe it's better to ignore the danger than draw attention to it, lest he risk jinxing it. He decides not to voice his concern.

"So yeah, that's pretty much it," Shree says optimistically. She takes a look around her at the points of interest in the living room below, summing up the tour in one brief gesture. She looks back at him. "What do you think?"

"I like it," he says, peering down at the sofas below. He thinks about all the good times he's about to have on those sofas, all the conversations — this is gonna be a fun year. He looks back into the brown eyes of the innocently freckled woman before him with an expression of visible excitement. "I think it'll be a good fit."

She smiles genuinely upon hearing this, her silver nose ring shining back at him. "Me too!" she agrees. Such a bubbly girl.

Cooper descends the ladder and retrieves his luggage, now toppled on the floor by the feet of the pink futon. He struggles up to the loft with it, hauling the bag to the top with an exhausting heave.

"What are you up to later?" she asks him.

"Looking for a job," he says, pulling the weight along the rickety walkway toward his room.

"Did you want to grab dinner or something?"

He rubs the back of his neck, waffling in his mind. "Uhh, I'm not sure. What time does Mel arrive?" he wonders aloud.

"Tomorrow afternoon."

"Oh, okay," he answers. "Do you think we can save the roommate bonding for then? I wanna get settled first, if it's cool with you."

Shree nods emphatically. "Sure, no problem! Just knock on my door if you need anything." Her warm smile effervesces in the glow of the open room.

He retreats into his bedroom and lets his bag fall to the floor. He considers the bare mattress lying in the middle of the floor. *Fully furnished three bedroom,* he thinks, remembering the online classified ad that brought him to where he stands now, a forty-square-foot chicken coop with a puke-stained bed on the floor. That's truth in advertising right there.

He pulls out his phone to check the time, only to see the battery level at 6%. In a panic, he grabs his charger from his backpack. He plugs the cord into the magic hole in the wall that provides heat and light and everything that makes his life what it is. Hell if he knows how it all works, just that it does. Instantly his device begins to draw electricity out of the wall and into its battery. Like magic.

He takes a seat on his bed and leans his back against the far wall. Out of his backpack he pulls his computer, opening its clamshell into his lap. He unlocks the screen with his finger, revealing an open internet browser on a job listings webpage. He highlights the text in the search bar and types "every level nyc." Enter.

2,800 results. He begins to sift.

Mel arrives the following afternoon, only minutes after Shree returns from church. In her hand is a leather suitcase full of clothes, a large bag of makeup, and her computer — the essentials. Shree gets the door and exchanges her customary salutations, taking Mel on the same brief tour she had given Cooper last night.

Hearing the commotion, Cooper emerges from his room to greet his new companion. The first thing he notices is how strikingly pretty she is, even more so than her profile let on — tall and blonde, with strong features and long, silky hair. Her squared jaw and imposing height signals to him that she's way out of his league, but he wouldn't make any advances anyway; that could make for a potentially awkward experience for everyone. But under the right circumstances, hey, you never know.

"So this is it, huh?" she asks, seeing her bedroom for the first time. Cooper edges past her and on down to the kitchen.

"Yep, our home for the next twelve months," says Shree.

The new arrival looks around approvingly. "We'll party it up," Mel promises. "You guys wanna buy some wine and chill in the living room tonight?"

Shree puts her hand on her chin ponderously. "Oh, umm, well it's Sunday night," she says. "It's my first day with students tomorrow."

Mel laughs dismissively. "Yeah but it's our first night together. We should celebrate!"

Shree pauses to consider it. "I guess I could go for some wine. It'll probably help soothe my nerves," she concedes. "I'm in." She leans over the wooden rail. "Hey Cooper!"

The boy pokes his head up from the refrigerator below. "Yeah?" he asks.

"You wanna have some drinks tonight?"

He squints uncertainly. "Well tomorrow's my first day of class," he begins. He sees Mel's mouth begin to open in protest, so he cuts her off before she has a chance to say anything: "But yeah, I'd be down."

"Alright then," Mel replies. "In the meantime, I'm gonna unpack. I'll see you guys later." She disappears into her room and drops her suitcase onto the bed, watching it bounce once before flopping onto the wooden floor.

Shree returns to her own adjacent room. Her pillow is propped upright against the wall for support. She leans back relaxingly, happy that her new roommates have arrived at last. She can hear Mel rustling about in the room beside her.

She grabs the shiny silver laptop off the floor beside her bed. The logo on its back alights with the awakening of its glossy, pixel-dense screen. She opens a web browser and navigates to a video subscription library. She selects a program and lets it play, settling down in her bed with an open box of cheese crackers by her side.

After making herself at home, Mel fetches some booze from a nearby supermarket. Having distributed the alcohol to her new roommates, she takes a long swig from her bottle of merlot as she leans back into the orange sofa cushion.

"So dudes," she says, "here we are, a bunch of young twenty somethings in New York City. We fucking made it." She holds the bottle high in the air.

"Big city living," Cooper muses, relishing the freedom of adulthood. "This is my first time living in an apartment."

"Oooh, big man now," Mel gibes. "No more RA to bust you for smoking in your room."

"What are you studying again?" asks Shree in between gulps of sweet white wine. Music plays in the background off Mel's computer.

"Business," he says flatly.

Mel burps. "Oh God, I remember those days," she says with a laugh.

"Were you a business major too?" asks Shree.

"Fuck no," Mel barks. "But I *did* minor in econ, and I endured my fair share of business classes."

"Yeah, it's no walk in the park," Cooper chuckles. "But it's almost over. One more year."

"Hell yeah," says Mel, bumping her bottle against Cooper's with a clink. "You better make it count."

"Senior year was definitely the best year of college for me," Shree relates. "I did so much stupid shit." She lets out a laugh as she probes memories of case races and midnight excursions to playgrounds. She pours a heaping drink into her wine glass. She sips it laboriously, contemplating every note.

"My God," she says, "tomorrow I'm gonna be teaching kids."

"That's so insane," says Mel.

"I know! I don't feel even the slightest bit ready." She takes another sip.

"What grade?" Cooper asks.

"Sixth. Science. I student taught with eighth graders, so I figure it won't be too different."

"Man, good luck with that," he says. "I got class first thing in the morning, blah."

"You should pregame it," Mel suggests. "It's the first day. It's not like anything important happens." She smiles at him teasingly.

He laughs mildly. "Yeah, maybe." He thinks it through for a second.

Perhaps a little alcohol wouldn't be such a bad idea, exactly; it might help him loosen up and talk to people, so he can start making friends early.

"He's thinking about it," Mel says with a laugh, watching the gears turn in Cooper's head.

"I'm seriously considering it now," he admits.

"Well let me know how it goes if you do," Mel jests.

"Wait, so what is it you do again?" asks Cooper.

Mel sips her bottle of wine. "I'm still looking for a job right now."

"Oh word, me too," Cooper says. "I have basically a month to figure it out before I run out of money."

"Yeah, my boyfriend is covering the bills for me right now," she says.

An hour passes. Shree reaches the end of her third prodigious glass of wine. Mel has long since finished her own bottle, and Cooper is working on his fourth beer. The new roommates are a lot louder than before, cranking the music streaming over the interwebs. The theme tonight is Katy Perry.

"Her music just *speaks* to me, ya know?" Mel proclaims. She's standing on the couch cushion that now sits in the middle of the floor. "She's just like, 'it's all hot and cold,' ya know? I feel like that's how the world is, just hot and cold, people say this and mean that. It's bullshit!" The young philosopher pauses for a moment to consider her thought, then, finding nothing, begins spinning on the cushion.

A deep belch erupts from Cooper's stomach. "I'll be honest with ya, Mel, I have no idea what you're talking about," he says. He points at her accusingly and winks. "But I like it."

Shree sways her head back and forth not quite to the rhythm of the music, eyes closed and wearing a grin. "I think I kinda get what you mean," she says without ceasing her personal groove. "Sometimes people are just like, they say they hate something, and then they *do that thing*! They're such, what's that word..." She thrusts her hand insistently into the air, searching.

"Well you know what I mean," she offers. "The word."

Mel leaps to her feet and lands on the cushion in a lotus. "Yeah, kinda like the douchebags that call girls sluts even though they're sluts too. You're not like them, are you Coop?"

Cooper looks around. "Me? Nah."

"Hypocrite!" Shree shouts. "Hypocrite is the word I was, I meant before. Those people are hypocrites."

"Yeah, exactly," Mel agrees.

"Wait, you said you have a boyfriend?" Shree asks her, as if she hasn't already seen the photos.

"Yes ma'am," Mel replies, pulling out her phone. She flicks on the screen, revealing a selfie of her and her beau, each making half a heart with their curled fingers.

"What about you, Coop?" Shree asks.

"Nah, no boyfriends," he says. "Looking for a girl this year though."

"Well New York is a great place to pick up girls," Mel reminds him.

"Yeah, I guess so," he says passively.

"Ahh this life is gonna be so good!" she declares, clapping her hands with glee. "Guys, we're all so *fucking* young. We're gonna kick so much ass."

Shree looks at her phone, checking the time. "Holy shit," she says, standing up with alarm. "It's almost midnight. I like." She stops speaking as her eyes glaze over. Her vision dissolves to purple, and she lurches back onto the couch.

Cooper looks at her distressingly. "Whoa, you alright?"

"Yeah, I think I got up too fast," she says with a laugh. "I gotta go to bed." She rises from the couch waist-first, her torso following until she stands fully erect. She hobbles toward the rope ladder.

"Yeah, I probably should get going too," Cooper agrees.

"You guys are so lame," Mel discourages.

"Umm, I have to teach students tomorrow morning and I'm way too drunk right now. What are you doing again?" Shree pesters.

"Sleeping till noon," Mel replies honestly. They both laugh.

Shree stands beside the ladder and stares up. The loft appears to be receding higher into the air, growing ever taller. She takes a deep breath, ready to battle gravity.

"Alright," she says aloud, "one step at a time here." She grabs a rung with her hands and holds it tight, raising a foot to meet the lowest step. One by one, she pulls herself up, Mel laughing from behind as she watches the struggle. Shree reaches the top and slowly meanders down the hall into her room. She pushes the door shut behind her.

Mel is still laughing at the girl's drunken maneuvering. A moment passes, and she hears a doorknob turn.

"Goodnight, roomies!" pipes the teacher through the crack in the door. She shuts it again and collapses on her bed.

"Night!" Mel shouts back, looking over to Cooper.

He sits in silence, wearing a wide smile on his face. He's overly tired and wants to be in his room, but something makes him think he should stay a little longer. And in his drunken stupor, he takes in the vista of his roommate: her bright, straight hair, her smooth white skin, her smoldering eyes with long lashes, and a tight white tank top revealing her exquisitely flat midriff. If it wasn't for the excess makeup, she would be a perfect ten.

He puts down his beer. "Well," he announces, "I guess I should get up."

"Alright," she replies, her eyes closed as she reclines on the floor. "You gonna make it up the ladder?"

"I think so," he says with boundless confidence. "I'm drunk, but I'm not that drunk." He waddles over to the ladder, looking back at the girl on the floor before he attempts the climb.

"Thanks for the company," he tells her. "This is gonna be a good year."

Mel smiles, her eyes still closed. "Oh, it will be," she says. "We're gonna have fun."

The Last Resort

The air on this crisp September morning is refreshingly cool, a dramatic contrast from the heat waves of midsummer that had made it such an unbearable few months in the city. The window is open in the living room, letting in the breeze with tones of autumn's chill.

Shree scurries about her comically tiny bedroom, trying not to make too much noise so as not to disturb her sleeping roommate on the other side of the partition. She picks up a stack of papers by her bedside that she had been reading in her drunken stupor before she resigned to sleep — a collection of lesson plans, notes she had scribbled from her days as a student teacher, and a cheat sheet of classroom rules that she printed off the internet. She stuffs them into a green folder and she puts it in her tote. She grabs a binder off the floor and throws it in as well.

She quickly changes into the clothes she had set aside for this day: a burgundy blouse and black skirt. She grabs her bag of makeup and places it among her school supplies. Then she opens the door and closes it silently behind her.

A cobalt blue sky greets her upon exiting the building. She takes a long breath and inhales the cool city air. A grin slides across her face as she steps onto the sidewalk. She begins her commute to school in the early light of morning.

After an hour-and-a-half train ride and several subsequent blocks of walking, Shree makes it to her school in the Bronx. She climbs the trodden stairs up to her classroom, arriving as early as she had planned for before she got wasted last night. She puts her tote on her plush teacher's chair to the back right corner of the room. This is *her* room, her domain.

She makes note of all the decorations she labored over all last week in prep for today: the large calendar, the frog-themed bookshelves, the student names written on different colored stars on the bulletin board. The gleaming whiteboard, the uncluttered table. Row after row of brown desks with their fake wood tabletop laminations, occupied only by small plastic chairs with tennis ball shoes. School. She's back in school. Only this time, she's in charge.

The new teacher takes her seat; class begins in two hours. Despite her fatigue, she's amped and ready. She begins reaching her fingers to unzip her tote when a visitor strides in unannounced. It's the principal.

"Sharifa," the tall woman says as she enters, calling Shree by her full name. "We have an issue."

Shree listens numbly as the principal explains how an unexpected influx of new students means she's absorbing an extra ten students in her class.

"*Ten* students?" she protests. "But I already have 26."

"Now you have 36," the principal states.

"Where am I going to put ten extra desks?" Shree asks desperately.

"Well, that's the bad news," the principal begins. "Safety code states that our classrooms can only accommodate 32 students at a time. The district is aware of our contingency here and has recommended we move you to a different room."

Shree looks around at her fully-furnished classroom, presently being taken from her without provocation or consultation. "But Principal Sanchez, my room is already set up," she cries.

"I'm sorry, Sharifa," Sanchez says gravely. "My hands are tied. We had no idea we were getting so many new students."

Shree sighs discouragingly. "Where am I moving to?" she asks heavily.

"The only space big enough is the gym," utters Sanchez. "We've already started moving desks and we're setting up a work space for you."

"The *gym?*" Shree squawks. "How am I supposed to teach in there?"

"We're working on it," is the unsatisfying reply.

"And gym class?"

"We're working on it."

Shree folds her arms unhappily. "This is ridiculous," she opines.

"It is," agrees the principal. "Now help me move some of these desks."

. . .

Meanwhile, 3000 kilometers to the southwest, two dark shapes make their approach to a gray-sided building. The compound is nondescript, its only identification the small logo on its side:

C

The agents don military-style desert camouflage fatigues, fully aware that the garments will do little to mask their approach from the sophisticated monitoring equipment installed around the enormous facility. But this doesn't even factor into their minds as they march toward the door, their enemy's cameras rendered irrelevant by the ingenious sabotage being worked by their leaders back in New York.

Both agents are dripping with technology. They have small headsets in their left ears, swinging down into a microphone in front of their mouths. The young woman is wearing an electronic device over her eyes, recording everything she sees through twin cameras on each lens. The glasses have a built-in antenna in the frame that transmits this data back to the vigilant eyes in the situation room back at Brooklyn Project headquarters.

Nevaeh Brooks sits at a desk in Williamsburg, watching the live feed from a large screen. Her father, Dr. Jamal Brooks, sits beside her. Two small touchscreens sit under the projected image on either side of the desk, both displaying information about their agents' exact position and biometric activity as recorded through the watches on their wrists.

"Calm it down, Iri," Nevaeh suggests. "You want to keep your heartrate low. That goes for you too, Tyrone."

"Yes, Sister," Iri says, trying to take her advice. She knows this is a one-way trip, but there's too much at stake for her to fuck it up now.

She and Tyrone step carefully across the dusty brown sand. Her watch glints in the desert sunlight as her hand hovers above the pistol slung over her right hip. It wouldn't be the first time she wound up in a firefight, and she prays that it will be her last, with all the uncertainty it entails.

From a tablet in Nevaeh's hand, the telemetry from the security cameras of the Coke compound play in tiles that fill quadrants of the screen. Each of them is being fed prerecorded video, captured by BKP in the week leading up to the raid now in progress. Through the mysterious connections of the darkweb, they have had access to critical data from this deadly Coke factory, including its security operations. Data-mining is big business these days, and for the right price, anything can be known. The software was hacked easily; it wasn't long before they were able to send fabricated data to the computer stations controlling the building's security, affording their agents a chance to walk in undetected.

Two beefy security men sit watch in the front of the building. A screen in front of them displays the same four camera feeds on Nevaeh's tablet. Nothing is out of the ordinary; there's no one as far as the eye can see, north, south, east and west. They sip coffee in silence.

Iri and Tyrone have reached the heavy steel door of the rear service entrance.

"Open it," she says into the headset as she draws her weapon.

Nevaeh obliges, pressing a button on her screen that sends a command to the security network of the compound. She watches the door open through Iri's eyes, holding her breath as the agent steps inside cautiously.

The garage is empty; aside from a few large wooden crates against the side, there isn't much to see, nor many places to hide. Tyrone follows behind her, carrying a black backpack with their mission-critical pwnie device stashed away inside. They move briskly across the empty concrete floor, reaching a door on the far side. It's unlocked. They open it silently and enter, guns drawn low.

The agents find themselves in a long hallway, terminating in an exit on one end and hanging a left on the other. They follow the bend, revealing rows of more doors, three to a side. They slink discreetly down the hall, keen not to make a noise. Although their extensive hacking revealed many important details, the building itself has no internal cameras, so there's no way to know how many people are in here, or what they're doing.

The guards continue to sit patiently behind the desk, their security screens displaying the building's climate-control metrics. Ventilation system is humming along at a constant 80% power. The temperature is an unwavering 20 degrees Celsius. The humidity is 44% — wait, the garage sensor just ticked down to 42. With nothing to indicate a malfunction, the men turn to each other curiously.

Nevaeh notices the change from her tablet, and a sharp chill runs down her spine. "Shit," she breathes.

"What is it," Tyrone whispers into his mouthpiece.

"Our feed isn't compensating for humidity," she reports. "Hopefully the Cokeheads don't think anything of a 2% increase in humidity. But keep your eyes peeled."

Iri motions to a door, deeming it their first choice; Tyrone nods. Her heart rate increases rapidly as she wraps her hand around the knob, the door creaking open.

It's an unlit storage closet, filled with various janitorial supplies: mop, broom, bucket, slop sink. She closes the door and they continue down the hall.

One of the security guards has left his desk, scratching his dark moustache as he makes his way to the garage. Nothing about the environment of this building ever makes a 2% change unless he tells it to. Something must be amiss.

Iri opens the second door on their right. Behind it is the mainframe room they've been searching for, replete with tall banks of whirring computer hardware, an infinite tangle of wires dribbling forth into power strips attached to the wall. Jackpot.

"Quick, Tyrone, swap out the pwnie," Nevaeh commands.

He hustles into the room, Iri closing the door behind him. They stand silently in the dim red glow of the room. Pushing a button on the side of his watch, Tyrone emits a bright light from his wrist, allowing him to see everything he aims his hand at.

He puts the backpack on the floor and fishes out the pwnie device. It looks like a standard power strip, but at its core it incorporates a secret, tiny computer, tucked unnoticed into the black casing. This machine, running powerful software designed by an elite hacker hired for the task, is the key to their snooping, the centerpiece of the mission that they've risked their lives to fulfill. He begins playing with the wires and cables in the room, looking for a port to hook up this expensive surveillance device.

The moustached guard enters the long hallway, passing right by the door to the mainframe room and on to the garage. Iri hears him walk by, putting her finger over her mouth to remind Tyrone to be absolutely silent.

Finally he's found a connection. He inserts the pwnie hookup and powers it on, booting its software and integrating it with the larger mainframe. The malicious code, a torrent of ones and zeros, rushes to meet the invisible barrier separating the more sensitive compartments of its hard drive from less-protected access points. At its gates stands a mammoth security field, prepared to flit away anything it finds suspicious. He hopes the tiny packet of necessary data will make it through without notice.

The code sneaks through; the heist is a success. Unlocking this gate, the pwnie gains full access to the computer network, providing an inside look at the true nature of this mysterious building in the heart of Mexico. The uplink is carried over a cellular network to Nevaeh's computer in New York.

"I have it," she tells them, her tablet screen opening a new program with the pwnie uplink. "Time for Plan B."

Tyrone nods, armed with the thumb drive impregnated with the data-mining virus. Now to find the proper port. There is none in the main-frame room; they'll have to keep searching.

Iri pokes her head out from behind the door. She hears faint footsteps moving away from them. They continue along the hallway.

The security guard has reached the garage. Seeing nothing but yesterday's aerosol delivery, he turns around, determined to uncover the reason behind this unexpected humidity change. He starts back down the hall.

"Right there," says Nevaeh, viewing Iri's feed. "Check behind that door. If there's nothing, then get the hell out." She bites her lip uncomfortably.

Her father watches the screen with unflagging focus. "You should order them back now," Jamal tells his eldest daughter. "We can't risk it."

"Yes we can," she says, muting the comlink. "If they find the pwnie, this will be our only glimpse into this place. Our people know what they're doing."

Iri opens the door. It gives way to a vast warehouse that occupies most of the building. Tall gray canisters stand inside, hooked up to an elaborate lattice of steel pipes and hoses. Whatever it is they're working on, it's huge.

She and Tyrone approach the canisters in awe, recognizing them as a peek into Hell. The tube nearest them reads "Sulfuric Acid 500 KG" in faded red ink across its dull metal sides. The hoses that tap the capsule lead into an enormous nozzle-like device in the middle of the room. It is ensconced in blue scaffolding, clearly still under construction. There are twelve huge containers hooked up to it, but there are slots in the lattice for over a hundred more.

"Are you seeing this?" Iri quavers into the microphone.

"Affirmative," Nevaeh states.

"Dirty Cokeheads."

Just then, the door flies open behind them. Two Mexican guards rush in, guns in hand.

Iri and Tyrone make a dash for cover. They dive behind the cans of aerosol, knowing the guards wouldn't shoot at them there, lest they risk the whole operation.

The guards shout in Español as they run after the intruders. An alarm claxon sounds; the building is coming alive. Iri and Tyrone are trapped, cover blown.

The heartrate monitor blares red on Nevaeh's screen: fight-or-flight is kicking in. The projector screen in front of her shows Iri's panic, the image shaking as her body racks with terror. The sound of her labored breathing fills the speakers loudly, and the shape of two pairs of feet can be seen approaching.

Jamal, his own tablet in hand, rewinds the video to when they first entered the room. He spies a door on the far wall; it may be an escape.

"Iri, turn to your 4 o'clock," he commands, pointing her in the appropriate direction.

The agent complies, noting the door in the corner and the red light flashing above it. The security guards are rounding from the opposite direction; maybe they can make a run for it.

Without giving herself even a second to think about it, she grabs Tyrone's hand and sprints for the door.

The guards emerge to see them charging for the exit. "*Detenen*!" one shouts to the scrambling agents. The moustached man fires his weapon; an ear-splitting noise rings out through the warehouse.

They're still running — the bullet drives right into the wall. Iri reaches the door, throwing her weight upon the crash bar. She clears the building and is hit by broad sunlight. She ducks behind the door as she holds the way open for her compatriot. They can still pull this off.

Two more shots peal out from the bloodthirsty guards. One of them pierces Tyrone in the leg; the second plunges deep into his back. He crumples onto his stomach with a wail, collapsing onto the sandy ground outside.

Iri pulls him clear of the door and shuts it. The guards are in pursuit, but Tyrone has been gravely injured. Panic lights through her mind — she has to act. With all her effort, she lifts her bleeding partner over her back and scampers around the side of the building.

The guards appear through the exit, their heads turning wildly in search for the intruders. Following the trail of blood, they gallop in Iri's direction.

She sprints madly toward the road where their contingency getaway truck awaits. She can hear the guards firing at her from behind, bullets whistling by. They're gaining slowly, but she's nearly to the truck. Her partner bleeds a steady trickle right down her back.

"Run, Iri!" Nevaeh shouts, standing nervously from her seat. It could be the last thing she ever says to her.

Iri reaches the pickup and slings her wounded in the back. She opens the door and clambers inside, the keys ready in the ignition.

Bullets impact the side of the vehicle with sharp thuds. She starts the engine and stomps her foot onto the gas, sending the truck forward promptly. It tears down the empty road, kicking up dust as a puddle of blood soaks into the sand.

Shots whir through the yellow vortex behind her, but the compound quickly disappears in the distance. It looks like she's in the clear. After a few moments, she pulls over to the side of the road.

"Iri!" Nevaeh shouts. "Keep driving!"

But the agent ignores her orders, instead climbing out and onto the flatbed to assess the condition her stricken friend. He's covered in dirt, a pool of his blood filling the bed of the pickup. He has two gaping holes in his body, red drooling out of his mouth. The look of terror on his face says it all. They both know he's gonna die. But at least she'll be there for him when he does.

She holds his hand in solidarity. "Hey," she says, peering into his eyes as he struggles with the pain. "We did it."

Tyrone breathes in weakly. He closes his eyes and smiles slightly, acknowledging the importance of his sacrifice. He hasn't the ability to speak; his senses are quickly draining.

Nevaeh bears witness to the scene from her projector screen. Her hands are over her face; she can't watch, only listen. Tyrone's heartrate monitor is pounding erratically — then, finally, it terminates into eternal silence. 0 beats per minute.

Iri feels his hand go limp in her grip. His trembling has ceased; his eyes remain closed. It's over. The peace is over.

She stands up alone, panting in exasperation. She looks down the road to the Coke compound. They could be after her. She needs to get moving.

She hoists Tyrone's body into the back seat. She takes her place behind the wheel, her hand dripping his blood over the furniture. With the rage of a gorilla, she feels a desire for retribution like she's never known before. She puts the truck into gear and slams on the accelerator, leaving streaks of red ink against the steering wheel. She speeds off into the distance, her fallen friend bouncing lifelessly behind her.

The climate wars have begun.

Life on Mars?

The people in the briefing room sit anxiously about the round table – except for Santi, the Clean Front delegate from LA, who is barking orders distressingly into a cellphone. There's a mood of grave despair on everyone's faces. They know there's nothing but bad news, as always.

There are only six present: Nevaeh and her father; two of her sisters, Jada and Tierra; Iri; and Santi. Together they're reviewing the data acquired from the raid, each navigating dynamic screens of information on their handhelds.

Iri had taken some high-res pictures from her optical headset, providing essential visual analysis of the geoweapon in the warehouse. Between this and the files sifting through the surveillance package, they may have enough intel to figure out their next move.

"Take a look at this," says Jada, pulling up a document from the Cokeheads' mainframe about aerosol shipments. "Twenty canisters of hydrogen sulfide for 'Project Chupacabra,'" she reads off the screen. Jamal taps his tablet to hers, allowing him to view the document on his device. He reads through it with panicked eyes.

"Jesus," he mutters soberly. "They're further along than we feared." His bony fingers pass up the screen to pull up another file, an invoice for massive quantities of sulphuric acid. "This design is way more elaborate than I imagined."

"They're deadly serious this time," Iri notes, ruing the loss of her friend and colleague. At least the casualty count wasn't two.

"They still need a delivery system," Nevaeh says, reviewing her tablet screen.

"Any airplane would do," Jamal reminds her. "They may have already modified it."

Santi's voice ceases its yell as he hangs up the phone, returning to the table in huffs. "My people are hitting up the darkweb for any cross leads," he tells them.

Jamal plugs some data into a complicated equation, mangled gibberish in letters and numbers that only experts like him can understand. Based on how much aerosol the Cokeheads have acquired, and how much their weapon appears to be designed to hold, he figures he can calculate the device's "yield" by climate modelling the effects of that much aerosol in the stratosphere. The numbers run; the algorithms crunch. They churn out a result.

The scientist stares at it disturbingly. He knows that this crude equation can't be as accurate as a full model which would take weeks to run, but this simple method is still a good estimate. And what it portends is disaster — a 3-degree Celsius decrease in global temperature, knocking the world back into the last ice age. He decides not to mention it aloud.

"If all this is correct," the scientist says, "and we have no reason to assume it isn't — then the Cokeheads are close to deploying this weapon." He looks up from the screen resolutely. "And I don't have to tell you the consequences if they do."

A long pause. Silence around the room.

"So what do we do?" asks Tierra.

"We act," Santi booms, a shell out of a cannon. "If we expose what they're up to, it could be the tipping point, the jumpstart to the revolution that we've been looking for." His hands gesticulate powerfully as he urges for immediate action. "If people find out about this weapon, public outrage would be further off the chart than last month's ocean data."

"We can't reveal what we know," Nevaeh counters. "That would be the end of our whole operation."

Santi's voice boils into a shout. "Well, like it or not, those Cokeheads have just fired the first shot of the climate wars, and Tyrone won't be the only one dead by the end of them. I'm not gonna let these junkies fuck up the climate."

"No one is gonna let them do that," Nevaeh dismisses. "Not as long as the Clean Front remains united. What I'm *saying* is that we should be patient and take time to think this through."

"He's right, Sister," Iri interjects. Nevaeh's eyes meet hers in surprise; the young soldier isn't afraid of saying it. "The climate wars are here. It's time to fight in them."

Nevaeh blinks at the counsel of her most trusted agent. She never expected to see her again. That the girl had volunteered for a suicide mission for their cause easily makes her one of the bravest people she's ever met.

"Two hundred thousand acres have already burned in California this year," Santi reminds the group. "The Keystone Fire still hasn't been contained, and it's only a few miles from my apartment. We've *taken* enough time. The clean revolution needs to happen now."

"I agree," adds Jamal in his deep, magnetizing voice. The women turn their heads to face their father. He motions to the tablets in everyone's hands. "We know what we need to know. It's time to move."

"I agree, dad," says Tierra, nodding solemnly.

They all look to Nevaeh. She's staring across the table into the distance, beyond the confines of their tiny room and out into the world at large. Her chin is pressed hard against her clasped hands. After what feels like an hour, her head moves up and down, slowly at first, then with more vigor, as she thinks it through and realizes they're right. The war has started. Now it's up to them to win it.

Nevaeh turns to her father; he flashes a knowing look. She looks to Iri, whose sullen eyes are turned down toward the table, still shell-shocked from her experiences. Finally she addresses the group. "Then we start recruitment," the leader announces.

At last they give in to their worst fears. Bloody insurrection. They didn't think it would come so soon. Like everyone, they thought there was more time. Always more time. Just not this time.

"I'm scheduling a Clean Front meeting for the beginning of next week," Nevaeh speaks. "I want everyone here."

She looks to Santi. "Tell your people to get ready," she orders. "This is it."

· · ·

Mel looks deep into her shot glass. The golden Tennessee whiskey inside frowns at her; she frowns back. After a brief staring contest, she grabs it off the table and empties it down her throat.

Shree is browsing her friend feed opposite the table, her laptop resting on her crossed legs as she reclines lengthwise down the pink futon. The scroll never ends.

Mel reviews the text message she had received a few hours earlier, the text that ruined her life to this point:

> Derek <3: this is why i cant do it anymore. im sorry but its time to move on. goodbye. 2:11pm

That final word, in its cold, sans-serif pixels, capped off a long exchange, a protracted breakup through the medium of digital text, leaving it feeling so unreal, so not-final.

"I can't believe he didn't have the balls to break up with me in person," Mel spits out. "Fucking coward."

"You don't need guys like that in your life," says Shree, not looking up from her screen. She's as many shots in as Mel.

"Or any guys," Mel thinks aloud.

"I know, right!" Shree agrees, turning to her roommate. "It's so hard to find a guy that isn't just a huge douche. This one guy I was texting about a month ago seemed pretty cool, but then he turned out to be a real creep and sent me a dick pic."

"Oh my god, Derek used to do that too!" Mel reveals. "Why the fuck do they send dick pics?"

"Cuz guys are pricks," Shree explains. "And they're obsessed with their pricks."

A notification blip chimes from her computer; she clicks on it immediately.

"What the— friend request from Danny Wegman?" she thinks aloud.

"Who's that?" asks Mel.

"A kid I knew in middle school," Shree says. She tilts her head in reminiscence of buried memories. "I literally haven't thought about him in 12 years. What's he friending me for?"

"Add him and find out," Mel says.

"What? Why?" Shree asks, not seeing the point. "That was, like, half a lifetime ago."

"Is he cute?"

Shree pauses for a moment to consider the small square picture next to his name. She squints to resolve more detail, but sees only pixels. Still, his smile is white, and his hair is sexy.

"Alright, he could be," she admits.

"Friend him."

Shree clicks the Add Friend button, and their two profiles connect. She surveils his feed for anything interesting. His relationship status is listed as Single.

"I guess the hunt begins again," Mel says, contemplating the single life. Watching her roommate perusing a potentially-attractive guy's page on the opposite couch fills Mel with envy, until, she realizes, she has an even better solution.

She grabs her phone off the couch and selects the pink and blue icon of her dating app. Her profile reactivates; she's gotten 300 new messages since she last logged on months ago. She would have more if she hadn't reached her inbox limit. She deletes all the messages without even reading them; she's ready to start fresh, and fresh means from right now.

"This guy's really hot, actually," Shree decides from his profile pictures.

Mel taps fingertips across her phone. "I just opened my dating app and already I've gotten a message," she says, so as not to feel left out from the male attention.

"Yeah right," Shree crows, calling her out. "You're not *that* pretty."

"On my profile I am," Mel retorts. "Look up melaniciouss92."

Shree complies, navigating to her page in a new browser tab. It loads a pink profile, reflecting Mel's identification as a woman. She looks flawless in all her pictures, but her page itself is sparse on details and text.

"You aren't 5'5"!" she says, referring to Mel's height. "You're like 5'9" at least!"

Mel laughs. "Yeah, but guys don't like it when you're *too* tall! They might filter me out if I put my actual height."

Shree closes the tab and returns to her feed. "You'll be catching dicks in no time," she assures.

Mel sighs. "I hope so." She begins to swipe through her options, left for no, right for yes. Derek <3 won't be easy to replace, she knows, but this will be a start. Her eye catches sight of the time at the top of her screen.

"It's getting late," she tells Shree. "I should be looking for a job." She opens the classifieds app on her phone and starts browsing. "Derek was gonna cover my rent this month, since I don't have enough in the bank. Now I don't know what to do."

Shree looks over at her struggling friend on the opposite couch. Mel's golden locks are scrunched into a ponytail, her violet shirt cut low. She's a girl that deserves a good relationship, if that's what she wants. Because Mel is a good person. And Shree trusts her.

"I'll cover rent for this month," she offers, prompting Mel to look up from her screen in bewilderment. "Don't worry about it."

She considers the offer with surprise. "Are you sure?"

Shree nods silently.

Touched by this show of compassion, Mel smiles tenderly at the friend sitting across from her. "Thank you," she purrs. "How'd you get to be so nice?"

Shree smiles, returning her attention to the screen. "Because I'm a Christian," she says.

. . .

Cooper sits in the lecture hall, tapping his pen against his binder as he listens to the professor with disinterest. His stomach roars achingly. He hasn't eaten since dinner last night; he can't afford to. For nearly a month he's been rationing his money, and with it his allotment of food. He wasn't prepared for this prolonged subsistence that he's endured since he arrived in the city; he thought he'd be employed within a week. The hunger is getting hard to ignore. He can feel his energy draining, his body weakening; malnutrition is setting in. He wakes up with aches, his stomach is always complaining, and he thinks with his appetite. If it isn't food, he isn't interested.

"It's the tragedy of the commons," the professor drones, "and that's why Malthus foresaw catastrophe."

Cooper's attention strays; he's heard all this before. With limited resources in mind, he begins sketching a simple tree onto his binder. He adds a fire burning from its canopy, then scribbles the entire thing out, leaving a dark square on the cover. No more tree.

"But Malthus was wrong," the professor qualifies. "The carrying capacity of the globe is much higher than he thought, because we developed fertilizers that allowed us to produce more food." The professor flips to the next page of his notes on the lectern. "We now have the ability to feed ten billion or more, in which case there would be no catastrophe."

A hand darts up from the middle of the room. "Excuse me," interrupts the student sitting two rows behind Cooper. Her words knife through the heavy atmosphere in the room.

Cooper spins round to see her. She's a small, dark-haired girl, her hand raised stubbornly in the air. Her large black eyes radiate intelligence and awareness, and her warm brown skin looks soft as cashmere to his eyes.

The professor acknowledges her comment, and she stands to be heard.

"What about carbon pollution from agriculture? How is that factored into the carrying capacity?" the girl asks.

The gentleman on stage smiles, removing his glasses thoughtfully to consider his student. "The young lady brings up an excellent point," the professor says. "Since we burn ten calories of fossil fuels to produce one calorie of food, the number of people we can feed is limited by the amount of carbon we can burn. And the answer, of course, is that none of this is factored into the equation, aside from the general idea of externalities."

The girl frowns at this. "I don't see how you can say ten billion people can live on this planet when we've eaten the whole world just feeding seven."

"I don't see why the city keeps investing money to fix the subway when it'll be underwater in sixty years anyway," the professor counters. "But that's what happens when externalities aren't factored in."

What the instructor says goes completely unnoticed by Cooper; his eyes are transfixed on the flawless beauty as she resumes her seat behind him. Her eyes make contact; Cooper, in a panic, looks away. But then he looks back, and smiles weakly while he still has her attention. Then, breaking the stare, he turns around again.

He keeps his eye on her as soon as class is dismissed. He follows her out the door, and when the moment seems right to initiate a conversation, he taps her on the shoulder.

"Hey, I liked your question in class," he says.

The young woman smiles receptively, her delicate face gleaming daintily. "Thanks. Someone had to say it."

"I'm Cooper by the way," he says, reaching out for a handshake.

"Natasha. Nice to meet you." She clasps his hand his hand with a grin.

"What year are you?"

"Senior," she answers. "This is my last semester actually. What about you?"

Cooper doesn't like hearing this; he was hoping to see her in more classes next term. "I'm a senior too," he tells her. "You got lunch plans?" he asks nervously.

She looks over to the boy, spare and genuine. "Actually, I'm heading to a meeting in about an hour," she explains. "I'm gonna eat there."

"Oh, that's cool," Cooper says, feigning interest as a dodge of her rejection. "What kind of meeting?"

"It's for the student decarb movement. I'm the president," she explains. "You're welcome to stop by if you want! It's open to everyone. Free food."

Cooper's ears prick up at the thought of a plentiful meal. "Oh yeah?" He checks his watch: 12:30pm. He's supposed to meet Mel for job-hunting in an hour.

"Damn, I actually can't go," he laments. "I'm doing something with my roommate."

Natasha smiles understandingly. "Ah, well that's okay. You can always come to the next one."

"Yeah," he agrees, recognizing an opportunity. "Actually, if I give you my number, do you think you could text me when the next one is?"

There's a glint in his eye that she recognizes as nothing more than a sneaky way to get her phone number, but she finds herself intrigued by his innocent forwardness all the same. "Yeah sure," she says, accepting this invitation. "I can do that."

"Cool," Cooper says as he pulls out his phone, hiding his excitement with suave nonchalance. "What's your number?"

Natasha gives him her ten digits. Cooper saves it to his contact list.

"Awesome," he replies. He did it. *Awesome.* "I'll text you."

"Alright, great," she says. "Anyway, I gotta get ready. I'll see you next week!"

Cooper watches her leave, feeling like a boss. That strategy almost never works.

His stomach growls again, more adamantly than before. With two dollars in his pocket, he wanders to a pizzeria that sells slices for a buck.

"Two, please," he tells the Spanish-speaking cashier, who grabs two paper plates and loads each with a slice.

He gazes affectionately at the pizza in his hands. He loves pizza, wants to marry pizza. Though the food on his plates amount to little more than cheesy wedges of bread, he knows they are packed with energy, brimming with life-saving calories, a full supply of petroleum for his empty gas tank. He gobbles them down savagely, swallowing them in less than a minute.

The hunger subsides, and with it the emergency. His body has been marginally appeased, at least for now. Through painful experience, Cooper is learning the primary motivation of adult life: the need to feed. Life is all about calories. That's ultimately what it boils down to. Food.

There's no need for a jacket on this late-September day. The sun's haze spreads widely overhead, sending temperatures into the 90s, cracking a hundred in the subways. Sweat collects in small pools underneath Cooper's armpits; he shouldn't have worn black. At least no one can see his pit stains.

He meets up with Mel a short time later. Walking through the city together, he realizes it's hard to find anything that hasn't been engineered by human hands. Every tree, bush and flower has been installed with purpose. A plastic six-pack ring blows down the street, buffeted by the hot wind.

A homeless man looks like he's about to expire in the midday sun. His body is slumped against the side of a building, catching the full intensity of the solar heat. A cardboard sign reads "Hungry and Sad" as it leans against his leg. A white cup with a few green bills sits by his waist. He doesn't look a day over thirty.

Cooper stops at the sight of him; Mel keeps walking by. After a moment of contemplation, he runs to catch up with her.

"He's our age!" Cooper shouts, aghast.

"Mhm," she replies.

"Shouldn't we help him?" he asks desperately.

"Sure, give him money," she allows.

"I don't have any cash."

"Then I guess you can't help him," she says.

Cooper looks back at the man as the distance between them increases. His heart sags in his chest. "I can't imagine being homeless at that age," he says, his face crumpled with dejection. "I thought homeless people were old."

"Maybe he's alcoholic," she says.

"So what?" he asks. "I know plenty of alcoholics. Pretty much everyone I meet in college is an alcoholic." He takes another glance at the young man on the sidewalk behind him. "Jesus, that could be me in six months."

"It probably will be," she says sarcastically.

"You're probably right," he agrees. "Why is that? Why is someone like him stuck on the streets and I'm not?"

"Because you're stuck with me instead."

Cooper frowns as he shifts his gaze forlornly into the carbon-choked sky. Its thick fuming miasma looks like the underside of a cotton blanket, wrapping the world in its warm embrace. The boy continues to sweat.

Meanwhile, half a world away, the waters off the coast of the Philippines churn with untold heat and power, a broiling vehemence the likes of which haven't been seen for a million years.

Die Young

Joseph is late to work.

His flame-colored pickup truck is stopped at a particularly long red light. Its thickly ribbed tires span two feet in diameter, each sporting thin black rims of carbon composite. Two horns curve menacingly toward the road from each side of the beast, and its swept back front lights give it a bloodthirsty look. It is a predator of the streets.

A pale blue hybrid idles at the stoplight in front of the truck. The puny car looks like a helpless bug, ready to be smashed flat by the grumbling monster behind it.

Joseph revs his engine in impatience. Exhaust spews out from the angry horns at the vehicle's head, letting out a roar. Unlike the white in his hair, his firebrand demeanor hasn't been tamed by age. He wants everyone to know that he's got places to be.

At last the light turns green. The hybrid doesn't react.

"Let's *go!*" Joseph shouts, slamming his fist into the center of his steering wheel. A peal of fury rings out from his truck.

The driver in front of him snaps to attention, returning his phone to the cupholder in the center console. He slowly accelerates through the intersection and up to speed.

This behavior insults Joseph, the other driver committing the ultimate felony: wasting someone else's time. Almost instinctively, he brings his hand to a cord that hangs from the ceiling. Gripping it firmly in his hand, he eyes the road ahead for an opportunity to pass. *There,* dead ahead, the street expands into two lanes. He readies his foot on the gas.

The engine coughs a blanket of smoke onto the road. As he pulls alongside, he slams on the accelerator, and gives the ceiling cord a swift downward rip. With the full power of diesel at its command, the truck responds to Joseph's control, racing past the puny blue car. As it does so, a thick sooty cloud of carbon is blown out the top exhaust vent, inundating the hybrid with gaseous sludge. The truck whips ahead like a bullet, leaving the hybrid to flounder in its heavy black contrail.

Joseph looks in his rearview mirror with delight at his plume of pollution, dense enough to obscure his victim.

Fucking greenie, he thinks.

The rest of his commute is free of incident. He pulls into a parking garage, where he receives his customary hat tip from the uniformed guard. Stepping outside into the unnatural warmth, his bright red tie phosphoresces in the morning light. In the distance, he can see the US Capitol, standing proud as ever outlined against the clear blue sky.

He makes his way down Constitution Avenue and into a white, L-shaped building with a marble façade. Without so much as a word to anyone, he passes through the security checkpoints and shuffles into his office. He closes the door behind him.

His mahogany desk is covered with organized papers, binders, and assorted business cards. His computer monitor is of an old design, and has been left on from the previous day, draining power overnight. Cases of bookshelves line the walls, most of them displaying framed pictures of him and his family. He grabs a thick stack of stapled papers and takes a seat in his cushioned leather chair. He reaches into his suit pocket and grabs his old man glasses, placing them on his nose.

"As you all know, large-scale deployment of renewable energy is unfeasible at the present time, and would come at an unacceptable cost to economic growth," he rehearses out loud. The vote on Representative

Hansen's emissions bill is scheduled for this afternoon; he intends for no yeas to be cast on the matter. "To implement these procedures now," he orates, "would be a death sentence to America."

A few raps from his door interrupt. It's his secretary.

"Good morning, Mr. Coke," she says, opening the door.

"Morning, Debra," he replies.

"You have a visitor."

From behind Debra's gray skirt, a child peeks at her grandfather seated at the desk. She's hiding a laugh with her hand, eager to surprise him.

"Who's there?" he entertains.

"It's meeeee!" comes the exuberant reply, the small child bounding into the well-appointed office. A paper flails in her hand.

"Well look who it is!" Joseph proclaims in delight, rising from his throne. "My favorite granddaughter."

She stops her charge in confusion. "I thought I was your only granddaughter!" she says.

"Of course you are!" Joseph says. "But can't you also be my favorite?" He reaches down to scoop her up, giving her a loving hug in the air.

Debra smiles as she walks out of the room, balanced on delicate high heels. A tall man takes her place in the doorframe; it's Joseph's son. He nods curtly at his father as the child returns to the floor.

"What are you doing here today, Molly?" Joseph asks the girl, crouching down to her level.

"I wanted to give you this," says the seven-year-old, holding up the paper with pride. It features a crude drawing in pastel crayon: a girl in a green dress holds colorful flowers in her hand, a bulbous yellow sun over her head. "It's me," she says.

Joseph marvels at the creation. "Why thank you, pumpkin. It's lovely. You made this yourself?"

Molly grins widely, her head shaking up and down.

"I know just where to put this," says her grandfather as he pulls himself up. He walks to one of his bookshelves, holding the drawing against it so that it faces his desk. He looks over to the man in the doorway and smiles.

"Hello, son," he says. The man walks in and they shake hands firmly. "Have you seen this?" Joseph asks him.

"Sure have. She's quite the artist these days," he replies, winking at his daughter. She beams back at him, revealing huge gaps in her smile.

"I thought you guys were flying out this morning," Joseph says.

"Her mother's got her this week, so she'll be flying back with her," his son explains. "But it's been a while since Molly's seen your office, and I wanted to spend a little more time with her before she left."

"Yes, well you caught me at an awkward time," says the old man. "I got to work a bit late today. Some flaming liberal can't tell a red light from a green one."

"Of course not," the younger gentleman replies. "I hope you rolled coal all over him."

Joseph smiles sadistically. "I'd say I blew him a healthy amount. Speaking of which, I'm getting ready for the big filibuster." He waves the thick stack of papers in the air. It's easily half a ream, maybe even an entire tree.

His son eyes the document. "For the carbon bill, right? You'll knock 'em dead. You always do." Molly wanders curiously around the room, running her tiny hands along everything she sees.

Joseph watches her entertainingly. "I sure hope so," he says. "Too much depends on it." He's thinking of the little girl.

The younger father responds with a knowing look, then turns to his daughter. "Wish grandpa good luck today, okay sweetie?"

Molly stops roving and looks up at her grandfather. "Good luck grandba!" she mispronounces, flashing the spaces in her smile.

Joseph pats her on the head. "Grandpa's gonna win for you," he assures her.

Just then a digital ringtone plays loudly, emanating from Joseph's pants. He pulls out the device, checking the caller ID on the screen. He recognizes it immediately. He swipes his thumb to the right without hesitation.

"Coke speaking," he answers tersely. He presses the phone hard against his ear, his eyes widening. "Where are you?" he asks. He turns his attention to the analog clock on the wall. "Alright, come on over." He hangs up and places it back in his pocket.

"I'm sorry everybody, I'm afraid I have to attend to some business," he tells his family.

Molly begins to pout. She folds her arms defiantly.

Seeing this, Joseph lowers himself downward until their eyes are at the same height. He smiles at her lovingly, then spreads his arms wide. Unable to resist the invitation, Molly smiles back and enters his embrace. He lifts her into the air again, giving her a tight squeeze. Her dangling legs kick wildly as she giggles with joy.

"Too hard!" she squeals.

"What was that?" Joseph asks, his grip unrelenting.

"Too hard!" she shouts between laughs.

He puts her down and rises. "Dad, make sure you both get home safe," he says to his son.

"You got it, grandba."

They share a stiff handshake before Joseph watches his progeny depart. Taking a seat, he returns to his filibuster speech.

"Regulating carbon is not such a priority as to curtail the growth of our economy," he continues reading out loud. "Americans have languished far too long under this restrictive administration..."

A short man enters the room, forgoing the formalities of a knock.

Joseph looks up from his script. "You got here fast," he says, tossing the fat document onto his desktop. He closes the door. "How bad is it?"

The man adjusts his glasses. "We're still working that out," he croaks. "But it looks like we have a bug."

. . .

"I pledge allegiance…" the children intone in unison to the voice over the intercom. Shree's forehead beads in the stuffy, cavernous gym.

"To the flag…" she mouths, watching as her kids stand at their desks, hands over their hearts.

"…of the United States of America…"

It's been a month since her forced relocation to the dimly-lit chamber, a refugee of the managerial incompetence that led to this unexpected student overpopulation.

"…And to the republic…"

Shree is still adjusting to the space. Her workspace is made of two student desks placed beside one another, her chair situated in such a way that her knees straddle the legs in the middle.

"…For which it stands…"

An industrial fan blows loudly in the background.

"…One nation…"

A student is sticking her tongue out at another girl two desks to her left. Shree ignores it.

"…Under God…"

"…Indivisible…"

"…with liberty and justice for all."

Done.

On her makeshift desk sits the new science textbooks that she received this morning, but she hasn't had a chance to review any of the lessons in depth — it's just like her student teaching days all over again. She knows she wasn't a *great* student teacher, but she always managed to wing it on minimal preparation, and this time she didn't have much of a choice: the books hadn't even arrived until today. Her students are already starting at a disadvantage for the state exams.

"I have new textbooks for each of you," she tells her class. She walks the rows of desks, placing a heavy volume in front of each child. As she approaches the back row, she hears a thump on the floor — one student has already dropped their shiny new book.

At another school in another part of the city, the stately Natasha twirls her parasol playfully as she and Cooper stroll along College Walk. The sun's intensity is like nothing Cooper's ever experienced in October before. Even in the burbs of Miami, it never felt this punishingly hot so late in the year.

"So you're heading to your environment club meeting?" he asks her as they travel to a coffee shop near campus.

"Yeah, you want to come?" she invites.

Cooper doesn't show much interest. If there's anything he cares less about than macroeconomics, it's the environment. But he would go just to spend more time with Natasha, if only he didn't have other priorities.

"Nah, I really have to focus on finding a job right now."

"At this rate you're never gonna come to a meeting," she bemoans.

"You never told me what it is that this club does, exactly," he tells her.

"We're trying to get Columbia to divest from fossil fuel companies," she explains. "We're planning a huge protest in the spring."

"Huh," mutters Cooper. "Why do you want them to divest?"

They arrive at the small shop on a city block corner, an austere little restaurant-slash-coffee-shop with a simple kitchen and plain wood tables. Its door reads "STRAUNT" in black capital letters. He holds the door open for her, proving that chivalry is not dead. He walks behind her into the line, a drop of sweat trickling down his leg.

"Because we have to show the carbon polluters that climate change is not acceptable," she tells him.

He squints at her puzzlingly, not sure what she means by this cryptic answer. "Climate change?" he asks. "You mean, like, global warming?"

She nods as the line advances. "Global warming, climate change, whatever you wanna call it."

Cooper receives a buzzing notification from his phone. He pulls it out of his pocket to check.

Mel: i got it 👍 1:39pm

He smiles at his screen and taps out a reply.

Cooper: Awesome, good work!! Now get me a job lol 1:39pm

He powers off his screen and returns it to his pocket. "My roommate just got a job," he muses. "Anyway, you were saying?"

"I'll have a grande," Natasha says to the barista now taking her order. She gives him her name; Cooper's next.

"I'll just take the cheapest cup of coffee you have," he says.

"Alright, so the mini-grande," says the cashier. "Do you want cream and sugar?"

"Is it extra?" Cooper asks.

The barista nods.

"Then no," he says. The receipt prints; Cooper doesn't tip, because every penny matters. He and Natasha stand by the counter awaiting their names.

"Here, let me just show you," Natasha says, pulling out her phone. She opens the video app and punches the phrase "bkp news" into the search bar. She selects a clip and tilts her phone into landscape mode, holding it up before him. Cooper watches with measured interest.

The video shows a reporter on the scene of a hurricane-battered coast. Palms flip wildly in the gale as a driving surf kicks up from the churning ocean. "As you can see, the waves are turning especially violent now, as the record-breaking Typhoon Powell makes landfall here in the Philippines," says the wind-whipped meteorologist.

The image cuts to an infographic of how warm the waters are around the Philippines, fueling the enormous storm that spins like a bladed top around its malevolent eye.

"Only time will tell how bad this turns out," the weatherwoman speaks, "but storms of this intensity are expected to increase in frequency dramatically in the next few decades as unmitigated global climate change continues unabated. More updates as we have them. I'm Tara McCandless, reporting for BKP News."

The short video fades to black, and Natasha returns her phone to her pocket.

"Wow," Cooper says. "That's terrible."

"Natasha! Cooper!" calls a voice from behind the counter. The two students grab their drinks and take a seat at a nearby table, hanging their purses off the sides of their chairs.

Cooper takes a big whiff of his brew, reveling in its intoxicating aroma. It transports him back to his youth.

"I don't know why," he says, "but whenever I smell coffee, I think of my mom. She used to always drink coffee."

Natasha stirs her coffee solemnly. "I wish I knew my mother," she says.

"Did you grow up without a mother too?" he asks impertinently; she nods. "Me too," he says, noting the connection.

She shrugs. "She died in childbirth," she says wistfully.

"Oh, I'm so sorry," Cooper laments. "That must have been really hard."

She takes a sip from her steaming cup. "It's easier to do without what you never had, I suppose. My father says she was the most enchanting person he'd ever known. All she wanted out of life was to have kids, and lots of them. She wanted to be a mother. And she wanted me to be a mother, apparently. She was looking forward to having grandkids. But she gave her life for mine."

She takes another large gulp of caffeine. "I'm just looking after her," she goes on. "My mother loved nature. That's why my dad married her, because he's a climate scientist, and she embodied everything he sought to protect. He has so many pictures of her with my sisters playing outside, and she honest to God looks like an angel. Like, I've never seen a prettier woman in my life. She was a flower."

Cooper smiles warmly, his heart softening like ice cream in the sun. "My mother loved being outside too. She used to play with me in the lake almost every day. But ever since she died I kinda stopped going outside for fun." He takes a swig of coffee. "I think I've just been mad that something as useless as a tornado took her away from me, and nature and I haven't really been cool since."

"Well maybe you should get cool with her," Natasha suggests. "Climate change is coming, and you may never have the chance to enjoy nature again."

"Huh," says Cooper. "Yeah, that seems like kinda a big deal."

"Umm yeah," she says facetiously. She stirs the black liquid in her cup. "Like, take this coffee for instance. The Arabica bean will be extinct before the end of the century, which means that our grandkids will never know what coffee tasted like."

Cooper tries to focus on the idea, but his vibrating phone provides a welcome distraction.

Mel: tonite we 🍾 🥂 1:49pm

He lets out a short laugh. "Looks like I'm getting drunk tonight," he tells her. "You got any plans?"

Natasha has pulled out her phone as well, texting with a furrowed brow. "Ugh," she grumbles.

"What's up?"

"I just…don't know. People are being stupid." She stands from her seat and grabs her purse off its back. "I'm sorry, I have to go take care of something."

Cooper feels uncomfortable watching her depart, but it's clear she has more important places to be, so he accepts it without protest. "Oh, alright then," he says, rising from his chair.

She grabs the coffee off the table. "Good luck on your job hunt today," she says, ready to turn for the door.

He opens his arms before she can turn away, ushering her into a hug. "Well it was nice seeing you," he says. They embrace tenderly, and she walks out of the restaurant with the grace of a ballerina, her almond skirt flowing elegantly as she traipses out the door.

Cooper returns his butt to his seat. He sips his coffee alone, staring at the liquid in the Styrofoam cup. Ground from beans, plucked from a tropical plant on the verge of extinction. How many billions of people have drank coffee in history? And within a hundred years, there won't be any more.

Better enjoy this cup, he thinks to himself. He drinks it down in long sips.

He hoists his tote over his shoulder and rises to his feet, casually tossing the empty cup into the trash on his way out. He notices a sign in the window as he approaches the door: "Now Hiring." Recognizing the opportunity, he turns around and walks up to the barista behind the counter.

"Hey, can I have a job application?"

The employee grabs a form from under the counter and slides it to Cooper. He jots down the relevant information and hands it back to her.

"Thanks," he says, turning for the door.

. . .

Early evening in Brooklyn. Mel and Shree are reclined on the orange couch, a tablet streaming music that fills the living room. They both have computers in their laps, browsing their friend feeds and checking their dating profiles. A margarita can rests loosely in Mel's grip. It's her second of the night; pregaming never gets old.

"Congrats again on the job, Mel. I guess that English degree went pretty far, didn't it?" asks Cooper, sitting across from them on the pink futon.

"Actually my biology degree," Mel says.

"Huh?" he asks. "I didn't know you have a bio degree."

Mel looks up from her computer. "I don't," she says with a wink. Her eyes return to the screen.

It takes Cooper a second to register her meaning. "You lied on your résumé?"

She doesn't look up from the glow. "It doesn't matter what degree I have really. The point is I can do the job."

Cooper frowns. "I don't think it's a good idea to lie to your new boss like that," he says.

"I know, it kills me to have to do it," says Mel, not altogether dishonestly. "But last week I got rejected from a fast food joint because they said I was 'overqualified.' So like, fuck recruiters."

Shree sighs loudly as she considers her dating profile. "So my date didn't go very well," she says.

"Yeah, what happened with that?" asks Mel.

"Well, like, first of all, he didn't look much like his profile picture. I think it was taken back in the early days of college or something, and he's like 26 now."

"Man, he's old," Cooper says.

"I know," Shree agrees. "Plus we just didn't seem to be a 90% match after all."

"So that's why you were home that night," Mel deduces, continuing to scroll through her feed.

Shree blushes. "I've never hooked up with someone I met online," she admits.

"Really? That's funny." Mel turns to face the fidgeting white boy. "What about you, Coop?"

He blinks with a dumb "What?" as if he had drifted away from the conversation.

"How many women have you slain in your boyhood?" she asks.

Cooper rubs his toe into the floor nervously. "Uhh, I'm a virgin, actually," he reveals. Shree looks up from her screen; Mel hardly moves at all.

"Not surprised," she says neutrally.

"Me neither, actually," says Shree.

"You need to get on that, dude."

He sighs. "I'm working on it," he says, thinking of Natasha.

"Well tonight's the night," Mel points out. "Parties are great opportunities to fuck."

He laughs nervously. "Yeah maybe, we'll see."

Shree gets up from her seat, closing her clamshell laptop. "Alright, I'm gonna go get ready," she says. Mel gets up without a word, following her roommate up the ladder.

Cooper ascends to his loft as well, expectantly giddy. Despite his trepidation, the possibility of hooking up with a girl is too irresistible to ignore. Like all 21-year-old males, most of his passions in life are driven by, above all, one engrossing preoccupation: to get laid.

But his wardrobe sucks. Look at these old shirts; they're so sophomore year. What can he wear that will make him look more like a bro and less like a pale nerd? He needs to look *frat.* Realizing he has no clothes that will fit the bill, he throws on a casual shirt and a pair of flat gray shorts to cover his lower half. There is nothing to distinguish his current appearance from that of any other day of his life; just looking at him, you'd never think he's been invited to a Brooklyn rooftop party.

But wait, he's still sober. He knows there's a bodega he can stop by on their way to the train, so maybe he can grab something there. But then he thinks. He puts his hand in his pocket to touch his wallet gingerly; he knows he only has eight dollars. The empty hole in his gut does the math quickly: that's eight slices of pizza, a whole pie in his pocket. Beer isn't food, he surmises, so he decides to leave the cheddar in his wallet for the chance to live another week.

Shree and Mel exit the bedroom with stunning elegance. Shree is clad in a blue lace club dress, terminating just above her knee. Mel is in a scarlet pencil dress which frills about her thighs, a marvel of sexual beauty. To Cooper, they are jewels in the sky, two of the coolest girls he's ever met. And here they all are, going out to their first party together. Living the young adult life in New York City.

Mel halts at the door as she takes out her phone, holding it high above her head. With her camera on selfie mode, she snaps a picture of Shree and herself posing in their irresistible sexiness. She posts the photo online, tagging Shree — and Cooper as well, who can be seen distantly in the background with a bewildered look on his face.

Both girls are now stopped on their phones, updating their cyber lives before they step outside, eager to preserve this moment in time in ones and zeros — the atoms of the digital universe. Shree's phone buzzes the notification of Mel's tag. She updates her status:

Shree Dunlap: Partyy nightttt ☺ – with Mel Livingston and Cooper Braydon
10:02pm

Business as Usual

"Shanelli!" Cooper shouts into the noisy room.

People stand impatiently, bags sulking under their eyes. Most of them are dressed for work, ready to tackle another day at the office. The rest are students, the few intrepid overachievers with the courage to attend the early-morning classes.

"Abel!"

The machines whir behind him as he operates his station with both hands. Being passed cup after fragrant cup, he lids them expertly, sending them along to their buyers.

"Tanner!"

The plastic cylinders are scribbled with customers' names, each of them eager to ingest the drug that powers their mornings.

It's Cooper's first day at work, and he still hasn't formally met everyone; they all got so caught up in the morning chaos that there wasn't any time for a real introduction. But the pre-class rush has subsided, and now seems a good opportunity.

"I'm Cooper, by the way," he says to the humans sharing the kitchen with him.

"I'm Nilas," says the stout, dark-skinned man who had been taking orders all morning.

Cooper's arm reaches forward in an uncertain trajectory; he can't tell if

Nilas is going for a handshake, a high five, or a fist bump. His hand fumbles into a fist as it collides awkwardly into Nilas's.

"I'm Divya," says a short woman as she looks up from her glowing phone. She has long curly hair and a white stud piercing through her left nostril. "It's nice to meet you."

Cooper bumps her fist, recognizing it as the easiest and least-confusing physical greeting in his arsenal.

Divya looks back at her phone screen. "I'm tagging you, Nilas," she says. The phone is so large that her small hands stretch uncomfortably just to hold it steady in her grip. Texting on that thing must be a nightmare.

"Word," the cashier responds, pulling out his own device. It buzzes with notification; he swipes down, opening his friend feed. He laughs at Divya's post and begins typing out a comment.

Cooper has questions for them, but he doesn't want to disturb them from their digital conversation. He decides to stock sugar packets instead.

"So I'm a Queens girl again," Divya says in Nilas's direction.

"Remind me again why you moved in with your mom at 24?" he asks.

"New York City rent," she moans.

Cooper, overhearing this, feels marginally better about his own knife-edge existence. Maybe leaving the nest really is as hard as it seems.

"I wish I had something to smoke right now," Nilas gripes.

Cooper seizes on this as a moment of potential conversation. "The bodega around the corner probably has cigarettes," he suggests.

Nilas looks at him confusingly. "Ew. What?" He shakes his head. "No way."

Cooper is confused. "I thought you said you wanted to smoke?"

"Yeah, not *cigarettes*," Nilas replies, as if it were obvious.

"Oh, sorry," says Cooper, withdrawing from the dialogue. He retreats to his phone screen, waiting desperately for the end of his shift. Seven hours to go.

. . .

Shree approaches her building with keys in hand. The air is sour with the smell of garbage baking in the sun; her bag is heavy from stacks of papers. It's entirely too hot for her taste, and the ride home has been longer than usual. All these factors compound like positive feedbacks into one altogether unhappy walk home.

Mel happens to be returning home from a bodega visit when she sees her roommate ahead of her. Recognizing an opportunity, she breaks into a sprint. Running silently, she bears on her roommate's position, pouncing on her back with both hands and then running off ahead. "Tag!" she says as she brakes to a halt.

Shree stops dead in her tracks, paralyzed with fear. "Jesus," she cries, "you scared the living shit out of me!" She has to catch her breath for a moment while Mel laughs at her genially. Shree can't help but laugh too, and they walk up to their room together, a couple of big kids coming home from the playground.

Shree plops down on the orange couch. "I have soooo many tests to grade," she whines. "Why did I decide to be a teacher?"

Mel makes a face at her. "I have no idea dude." She pops open a beer from the fridge, hands it to Shree, and grabs one for herself. "Let's just chill."

They sit on their respective couches and relax. Shree pulls out her laptop; Mel is texting on her phone.

"So I told my friend Marquis that he should come over while he's in town," Mel says between swigs of her beer. More texting, a hunger pang. Before she knows it, she's sleepwalking to the kitchen, reaching for a box on the upper shelf of the cabinet.

"What are you making?" Shree asks.

"Ramen. What else?" She retrieves a bag of noodles.

"This isn't college, dumbass," Shree taunts. "Cook some real food, like hot dogs or something."

"Don't tell me how to live my life," Mel spits.

Shree finishes her beer and closes her laptop. Returning it to her bag, she throws the heavy sack over her shoulder, and takes the whole thing up to her bedroom, fighting gravity every inch of the way.

Cooper gets home from his first day of work not long afterward. He climbs to his bedroom, thoroughly exhausted; it's been an endlessly hectic day. But after a month of living in abject poverty, at least he's earning bread again.

He pulls out his phone and selects the banking app to check his current balance: 886 dollars, split between savings and checking. Tomorrow is rent day; he also has a credit card payment due on Friday. He won't get his first paycheck for another two weeks. This might be cutting it close.

He opens the calculator app, and he takes out his wallet to see what's inside: three singles. He punches the numbers. 886 + 3. There's a ten-dollar tip waiting for him back at Straunt, which he adds to the equation. Then he subtracts his minimum credit card payment, followed by rent, and finally last month's utilities payment. He hits equals.

-54.

A wave of heat flashes over his forehead. He stares at the screen. *No,* he thinks; *that's not right.*

He does the math again. 886 + 3 + 10 – 31 – 900 – 22. Equals.

-54.

NOT ENOUGH, his mind blares. His heart pulses rapidly. He's gonna starve.

He imagines himself a peasant farmer in a dust bowl watching his crops succumb to drought, realizing that despite all his hard work, he's gonna go hungry this winter, and maybe even die.

-54.

The math is irrefutable. He'd be a fool to deny it; his survival depends on heeding this warning. Being out in the real world has suddenly become a horrifying experience. In his hand-to-mouth existence, he's never more than one bad mistake from death.

He pushes more buttons on his screen. He subtracts 35 dollars for the inevitable overdraft fee: -89. Factoring in a few dollars for buffer, he decides to call it an even hundred.

With mortal humility, he pulls out his phone and composes a text to his father:

Hey. So remember when I said I wouldn't need any financial help this year? I think I was wrong

He reviews the text before hitting send. His dad is gonna be so pissed. Cooper *promised* him that he'd be able to handle it from here, that he'd navigate this year like a grown man and not have to rely on dad's wallet for help. But here he is, asking for money.

He decides to rewrite the text.

I just got robbed by a some guy with a gun. He took my wallet. I had a hundred dollars in it. Do you think you can spot me and I'll pay you back?

Nah, that's rubbish. Plus there's a typo. There would be too many questions, and he'd probably have to cancel his bank cards before his dad believes him. He takes one more stab at it.

Hey dad! I just started my new job, but it's gonna be a couple weeks before I get paid. Do you think you could spot me a few bucks to help me get by for rent this month? I'll pay you back asap

Good enough. He hits Send, and watches the word bubble appear in the conversation thread. Now he waits. Maybe the reply won't be as harsh as he fears. After all, would his father really just abandon his kid when he's in need?

A buzz from the phone.

Dad: How much 5:18pm

Cooper winces. He types in the letters with trembling fingers.

Cooper: Just a hundred. 5:18pm

It's not an altogether unreasonable request. Still, he wants to make his father proud, to overcome the distance between them. And maybe, by granting him this small loan, it will draw them together, father helping offspring, dad and son.

Another buzz.

Dad: Sorry kiddo. You promised. 5:19pm

Cooper's eyes brighten with fear. He can't believe it. Not even a hundred bucks? The rejection is crushing, like a wrecking ball landing on his chest. *Fuck you, dad,* he thinks angrily.

He powers off the screen and throws his phone onto the couch beside him. What's he gonna do now? His eyes scan the room, until he focuses on the loft above. With nowhere else to turn, he decides to text his roommate.

Cooper: You home? 5:21pm
Shree: Yeah. 5:21pm

He lunges his frail body off the bed and walks over to her room, knocking twice on the door.

"Come in," says the voice.

Cooper stumbles inside. "I could use your help."

Shree is grading papers in bed; she looks up from her stack to consider the baby-faced white boy in her doorway. His cheeks are tender with deference, and his eyes look sad.

"Sure, Coop," she says, putting down an exam. "What's up?"

He rubs his hand up and down his forearm nervously. "I think I'm gonna be a hundred dollars short for rent this month," he explains. "Is there any way you can spot me until I get my first paycheck?"

She eyes him severely. If it was anything other than money he was asking for, she might not feel so personal about it. But her roommate needs her help, and ultimately that's all that matters. She sighs, checking her bank account from her phone.

"Between you and Mel, I'm gonna have nothing," she muses.

Cooper continues to rub his arm uncomfortably, not sure what else to say.

"Alright, a hundred dollars," she agrees.

His face lights up and he gives her a hug in bed. "Thank you," he says as they embrace, happy to have someone looking out for him. "I'll pay you back before the end of the month."

"Or else what?" she asks.

Cooper decides not to answer, changing the subject instead. "So what's all this?" he asks, pointing to the papers on her bed.

"Science tests," says Shree. "They take forever to grade, especially when these kids get every other answer wrong."

. . .

The sky is a golden orange as the sun reaches its lowest point on the horizon. People are out in droves with the relief from daytime temperatures, returning their parasols to their handbags and enjoying the cooler twilight breeze. Cooper, now through with work, walks up the gray stoop to his apartment.

He hears Mel's laugh when he walks inside, and he sees her sitting on the couch with some random guy, a few empty beer bottles littered about the cement floor. Shree is on the opposite couch; they're all watching some cartoon from her laptop, the fan on the floor blowing air softly over their sweaty bodies.

"Whatcha guys watching?" Cooper asks the party.

"Adventure Time," says Mel. "By the way, this is Marquis."

"Hey man," the stranger says, holding out a hand. Finally, a handshake. Cooper knows how to handle those.

But he was wrong; what Marquis intended was actually a complicated dap, which Cooper botches embarrassingly. He should have known better — Marquis looks way cooler than a simple handshake, with his large gages and top knot on his head.

"Did you just come from work?" Shree asks him.

"Yep, exhausting," he says. "I'm over it already."

She scoffs. "Pfff. Tell me about it. I'm like, over adulthood already."

Mel takes a sip from her beer and raises the bottle into the air. "I second that," she declares. "Let's go back to college."

Marquis breaks out a small bag of marijuana from his pocket, and begins rolling it into a joint on the black table. "You guys wanna smoke?" he asks.

"I thought you'd never ask," Mel groans.

"Shree? Cooper?" Marquis looks over his shoulder to them.

Shree shakes her head. "I'm good thanks." Cooper follows suit.

Just then Marquis gets a notification; he pulls out his phone to answer it. After a minute, an impatient Mel has to remind him to focus.

"Hey, get off your phone and roll that J!" she commands.

"Sorry, sorry. I'm trying to set up this date with my friend from college," he explains.

"Ooh, is he cute?"

"Take a look," Marquis says, showing the group his friend's profile picture.

Mel takes the phone and starts flipping through the stranger's photos, including a beach picture featuring washboard abs. "Oh daaaamn boy, hit that!" she teases.

"I intend to," Marquis says, trading the white paper roll for his phone. "What should my status be?" he asks Mel.

"I don't care," she replies, handing him the joint. "Your hit."

He quickly taps something out and powers off his screen. Then he takes a long draw from the joint and blows the exhaust straight into the fan, filling the room with the potent aroma of ganja.

"So what are you guys doing all night?" asks Shree.

"This," says Mel. "Probably get some pizza."

"It's Friday night," Shree reminds her, "you guys aren't going out?"

"I'm broke as shit," says Marquis. "Besides, I haven't smoked with my Mellie in way too long."

"It's true. This is our thing," Mel says.

Shree frowns. "Well I wanna go out," she says, turning toward Cooper. "What do you think, Coop, you wanna hit the town?"

He ponders it for a second. He's pretty tired from a long day, but he doesn't have anything better to do, and he just got paid.

"Yeah, okay. Why not."

The two millennials walk into a dive bar in the heart of Brooklyn; Shree flags the bartender with authority. "I'll take that," she says, pointing to the deal scribbled in chalk on the board behind him. He nods curtly and grabs an empty pitcher, filling it to the top with fizzy yellow beer.

"That's the way to do it," she hears someone say; it came from the man sitting two stools away. A half-full pitcher of amber sits in front of him. He turns to face them and smiles, holding up his beer in greeting.

Shree, recognizing the gesture, clinks pitchers with the man. "To cheap beer," she says.

The stranger smiles at her intently, entranced by her inviting brown eyes. "You guys are starting off the night right," he says to her. "What's your name?"

She looks over to Cooper, waving him over. She turns back to face the gentleman, deciding to sit beside him. A wide rectangular bandage covers the left side of his jaw, and his scraggly black hair droops into two knots on either side of his face. He looks homeless, and not much older than she is, if he's even older than her at all.

"I'm Shree," she says, meeting his gaze. His eyes are a stunning battleship gray, two sterling jewels in the darkened barroom.

The man smiles. "I'm Pando. It's a pleasure to meet you."

Cooper sits at the bar beside Shree, grabbing the pitcher out of her hand and pouring himself a glass. He replaces it in front of her.

"That's a cool name," Cooper says to the stranger. "Where are you from, Pando?"

Pando stares deeply into his pitcher. "Kabubwason," he mutters.

Cooper's ears prick up. He's heard that word before, but can't remember where or when.

"Where's that?" asks Shree, pouring herself a beer.

Pando's eyes shift to her. There's a marked beauty to him, his ragged appearance spellbinding in its brute sincerity. He looks angsty, broken, like an Asian Kurt Cobain. Charisma surrounds him like a magnetic field, grizzled and roguish, like the kind of person with a bunch of innate skills, a young MacGyver. He looks like he rolls his own cigarettes, and wedges open bottlecaps with buckknives. He looks like someone important.

"It was a city in the Philippines," he reveals, bringing the pitcher to his lips for a deep quaff.

That's where I heard it, Cooper remembers. *The typhoon.*

"It was?" Shree asks in confusion.

Pando nods grimly. "When Typhoon Powell blew through a few weeks ago, the storm surge washed it clean into the sea. There isn't much left now."

Shree's eyes widen with terror. "Oh my God, were you there?"

The man touches the bandage over his face. "Yeah."

She covers her mouth with her hand. "That's awful," she says. "I'm so sorry."

Pando shakes his head. "I lost my brother," he says painfully. "My mother stayed behind to help with the recovery. But I couldn't stay; there aren't enough beds."

"Holy shit," Cooper says, trying to involve himself. "So what are you doing here?"

Pando looks into the frothy amber in his glass. "Trying to get wasted," he says. "I've been couch surfing since I got back to America, but I'm getting kicked out tomorrow. I was only supposed to stay for two weeks."

"You're American?" Shree asks.

The man nods. "I was just visiting my family for a few months before I came back to the States. But then with the storm and everything..." He trails off, rotating the pitcher back and forth with his hands. He pulls it up to his face, drawing a mouthful of bitter alcohol down into his throat.

Shree and Cooper stare at him with concern.

"I just happened to be in the wrong place at the wrong time," Pando finishes. "Kind of like all of us. The wrong place at the wrong time."

Cooper nods in agreement, not catching any larger meaning behind his words.

"At least you lovely people are here with me, though," Pando says to them.

Shree smiles, flattered by the uninhibited openness of his remarks. "Well it's nice to meet you, Pando," she says.

"I'm Cooper, by the way," comes a quiet voice.

Pando extends him a fist. "Nice to meet you, man," he says with a bump. "So what do you guys do?"

"I'm a barista," Cooper says flatly.

Shree sips her beer and puts it on the counter. "Teacher."

Pando raises his glass in approval. "Right on, teachers are great," he enthuses. "What made you want to be a teacher?"

She grimaces, fumbling with her beer glass. "I just really love kids," she says. "I care about them a lot. I want to make their lives better."

After covering their résumés, the acquaintances move on to other topics. Pando finishes his pitcher and calls the bartender for another. The roommates work their way through theirs, quickly catching up. The conversation flows freely for over an hour, until the wounded man announces he should be heading back home.

"I think I'm ready to go, too," Shree agrees. "How are you feeling, Coop?"

Her roommate nods in agreement. He's had more than enough alcohol for one night; really, he just wants a sandwich.

Pando pays his tab, leaving a bill on the countertop for the bartender. Shree pulls out a plastic card from her phone wallet.

"I got it," she says to Cooper, assuring him that he won't have to spend his food money on this night of empty calories.

The three of them step into the late November air, not as biting as they remember it. They stride down the quiet sidewalk three-abreast, passing by rusted grates and broken mailboxes along the dilapidated street.

"So what's your plan?" Cooper asks Pando. "Are you gonna couch surf somewhere else?"

The man shrugs. "I haven't figured that out yet." He touches his hand to his bandaged cheek tenderly. "I'm more of a go-with-the-flow kind of guy."

Shree is worried by the sound of this. Though she just met this guy, his forthright honesty has been a refreshing change from the banality of most barroom encounters, when people shoot the shit over celebrities, television, and the latest internet memes, all in a vapid bid to get laid. Pando seems to be above that; intelligent, thoughtful, vulnerable. And she finds him attractive like no one else she's met before.

"Well if you want," she says, "you can crash on our couch until you get yourself situated."

Cooper looks over at her with furrowed brow — it's quite an offer to make without even asking her roommates. But when he thinks about it, he realizes he wouldn't mind. Mel, however, may be another story entirely.

"Really?" asks Pando, his silver eyes piercing the night.

She smiles politely. This man needs help; she'll be the one to provide it.

He beams enthusiastically. "What's your number?" he asks her.

Shree gives it to him, watching him write it down on a piece of scrap paper pulled from his pocket.

"Alright, I'll let you know!" the man enthuses. He opens his arms wide for a hug, which Shree is all too happy to oblige.

"Yeah, hit me up!" she says optimistically.

Pando holds out a fist for Cooper; they pound like old friends.

"I'm glad I met you guys tonight," Pando says. "I'll see you around."

Belle Glade Missionaries

Cooper opens his sleepy eyes with a yawn. He stretches his arms wide above his head, extending every muscle to full length, striking life and activity into his dormant body. After rubbing the drowsiness out of his eyes, he hauls himself up from his mattress and shuffles out his tiny bedroom in nothing but underwear.

With tired reflexes, he makes his way down the ladder and into the bathroom. He lazily reaches for the red toothbrush in the cup on the sink counter and runs it under a blast of hot water. He grabs the curly tube of toothpaste rolled up in the medicine cabinet, squeezing a line of glittering green jelly onto the plastic bristles. He sticks it in his mouth, running it over his teeth in small circles. After about a minute, he spits out the white foam and cups his hands under the faucet for a rinse.

He looks at himself in the mirror, passing his fingers tenderly over the shape of his chin. Not a single hair.

21 years old, you'd think I'd have more to show for it, he ponders. Other people his age are sporting thick beards, yet he barely even has peach fuzz.

He slogs back up to his bed and checks his phone out of habit; no texts. He opens the messaging app and scrolls to find Natasha's thread. His thumbs tap the keyboard.

Cooper: Hey! What are you up to on this lovely Saturday morning? 10:19am

He hits send, his text bubble appearing in the conversation. Almost instantly, the read report icon shuffles to the bottom of the screen: Natasha has read his text. Cooper keeps his eyes on the screen, waiting for her reply. But nothing happens, not even an animated ellipsis to indicate she's typing. He waits longer, trying not to look too much into it.

The seconds drag on like hours, the screen dimming from inactivity. Cooper taps the display to indicate he's still paying attention. *Maybe she didn't actually read it?* he wonders. He clicks on the pair of eyes at the bottom of the screen. "Read at 10:19am," it says.

She's probably busy, he thinks. But still he stares at the screen, hoping a message will come through. The screen dims again, and this time he lets it dip to black. His heart sinks. It feels like he's never gonna get close to this girl.

He reclines in his bed, disregarding Natasha's non-reply with a deep sigh. His eyes turn to the blue box of condoms under his nightstand, which has sat there untouched since he moved in. He's starting to wonder if he'll ever have the chance to use them.

Curious, he reaches over to the paper box; inside are eight plus two bonus rubbers, each neatly wrapped. He's never tried one on before. He figures he might as well know what it's like.

He pulls off his plaid boxers and grabs one of the blue squares out of the box, tearing it open. He delicately slips out the greasy piece of clear latex and flicks the wrapper onto the floor. Just the thought of putting on a condom has induced a sizeable boner, and he begins to fantasize about his economics class infatuation. As his thoughts turn filthier, he nurses his erection to full size. Sufficiently horny, he puts the condom on, sliding its curled sides down his throbbing penis.

He runs his finger up and down his shaft, feeling the smooth latex under his fingertips. With his thoughts straying to his sexual fantasy, he unconsciously wraps his hand around his dick, rubbing up and down, slowly at first, then faster and faster.

"Cooper," he hears Shree call, rapping loudly on the door.

He releases his grip and sits up in bed. "Yeah?"

"I just wanted to say—"

"What?" he interrupts. He can't hear her through the door.

Shree turns the handle, pushing the door ajar. "I *said* that I just—"

"Whoa, whoa, I didn't say come in!" he protests angrily as a flash of sweat pulses over him.

She catches a glimpse of Cooper's white, naked body on his bed. "Oh shit, sorry," she says, quickly closing the door.

Cooper's face sears with embarrassment.

"I *just wanted to say*," she calls out, "I got a text from Pando; he'll be coming over later."

"Oh. Okay." He hears Shree move away from the door and head downstairs; his boner is gone. He pulls the condom off and replaces his underwear. Acting like nothing happened, he grabs his computer off the floor.

He opens a new tab in his browser and navigates to a news feed, not the social media kind, but the kind with actual content reflecting actual news. One of the headlines is about Typhoon Powell. Thinking of the mysterious young man he met at the bar last night, he clicks it to see what it says.

The top of the page depicts two children holding hands as they stare heartbroken at an expanse of rubble and splintered homes. "Twins overlook the destruction of their family home in Kabubwason," the caption reads. The sight is eerily reminiscent of what he himself witnessed fifteen years ago.

The report describes in detail the hardship the victims have endured since the massive storm swept over the island city. Food and water rations are growing low, and the shortage of medical supplies has resulted in countless excess deaths. There are so many bodies that the locals have taken to dumping them out to sea, where they collect in vast rafts of human corpses that decompose in the boiling ocean. And all of this is happening as another storm forms not far from where Powell began its fearsome march to land.

Pando arrives later that morning, carrying with him nothing more than a small backpack slung over his shoulder. Shree is there to greet him.

"Hey," he says with a modest smile as she opens the door.

"Hi there!" she says fervently, taking him aback with an unexpected hug. Her cheek grazes the bandage on his face as she lets go. She smiles brightly. "Come on in!"

The man ambles slowly inside, glancing around as he enters his new temporary residence. The amenities are somewhat less spectacular than he anticipated, but being a climate refugee means taking whatever he can get.

She follows behind with uninhibited enthusiasm, her heart thumping with commotion. She's eager to show him around, not that there's much to show.

"So here's the kitchen..." she begins, gesturing to her right.

"Yes, I see," says Pando. His eyes dance around. He spies the living room in front of him, and begins making his way to it.

"Would you like a glass of water?" Shree asks.

"No thank you," he says politely. His only desire is a horizontal couch. He hasn't gotten a good rest in weeks; the sleep deficit controls every movement in his body as he straggles farther inside.

"Alright," she says. She skirts in front of him, leading him to the dual-couch setup in the middle of the chamber. "You can make yourself at home here."

Pando nods silently as he sleepwalks to the orange couch, judging it to be cozier than the pink futon. He drops his backpack by its armrest and collapses onto the plush cushions with a satisfying "Ahhh." He lays there face down for a moment.

She smiles gleefully. "Comfy?" she asks.

He turns his eye to look at her. "You have no idea," he says. "I haven't felt anything this soft in almost a month."

Her heart fills, blessed to be helping someone in need. That he happens to be hot is only a secondary motivation, but a strong secondary. "Is there anything I can get you?" she fawns.

He shakes his head sleepily. "I'm good, Shree, thank you so much." He rolls onto his feet, rising upright to a full foot above her head. He looks deliberately into her eyes, communicating his appreciation. "It's really nice of you to let me in like this."

She blushes, her heart twitching as the affection pours over her. She gazes back into his platinum eyes. She opens her mouth to speak, but no words come out. She hiccups instead.

Pando chuckles quietly. "So I haven't exactly slept yet today," he begins.

"Goodness!"

"Yeah. I know I just got here and all, but do you mind if I take a nap?"

"Of course not!" Shree emotes. "The couch is yours. Do you want me to grab you a blanket or something?" she asks dotingly.

"I think I'll be alright like this," he says, sitting down in his new home.

"Okay, well let me know if you need anything! My room is right up there," she says, pointing up to the leftmost door in the loft.

"Alright," says Pando, laying his body out on the fluffy cushions. He makes no further noise.

Shree stares at him for a moment, pondering the wounded young man. His shorts are smudged with gray soot, and his legs feature small cuts and scrapes that are mostly healed up. His black hair covers his eyes and nose, but the bandage is still visible on his cheek.

"By the way," she starts, but Pando is already asleep. His right arm dangles off the side of the couch at an awkward angle. Amused, she walks over to his side and gently replaces it on his stomach.

The sun has set by the time Mel returns to the apartment. As soon as she opens the door, she notes the human occupying the couch. Interested in what he looks like, she goes over to the dormant man, and is somewhat aghast at what she sees. He's still wearing his shoes, which look like they've walked a hundred kilometers through mud, swamp and glass. He looks like a homeless man. In some respect, she supposes, he is.

"Hmm," she mumbles as she heads to the kitchen. She opens the freezer and selects a prepackaged meal. She tears open the paper and removes the plastic tray of food. Using a fork, she garishly stabs the crinkly wrap, creating vents for steam to escape. She puts her frozen meal into the microwave and punches the door shut. Inputting a three-minute timer, she cooks the meal on high. A droning sound hums from the wondrous appliance.

Pando stirs at the noise from the kitchen. His blurry eyes reveal the shape of someone standing by the counter. He also notices his shoes are still on. He yawns and lifts himself languidly upright. He slips his feet out from his sneakers and gets up to greet this new person.

Mel hasn't noticed his movements; she's having a text conversation with one of her friends. She holds her phone up to her face to send a selfie. Out of the corner of the screen, she discerns a man walking towards her.

"Hey there," says Pando gently.

She screams as she turns round, startled by his salutation. "Oh my God, I just had a mini-heart attack," she says, only half joking.

"Oh, sorry," he says neutrally.

She takes a moment to catch her breath. "It's fine."

"I'm Pando," he says, offering his hand.

She notices the dirt under his fingernails, but can't refuse a handshake. "I've heard," she replies, meeting her hand with his. They shake weakly before releasing. "I'm Mel."

Shree has overheard Mel's shriek from the loft. She puts aside the papers she's grading and opens her door cautiously, sticking her head out to look below.

"What time is it?" asks Pando groggily.

Mel checks her phone. "8:30."

"Really? Daaaamn," he says. He goes back over to the couch to grab his knapsack, hoisting it onto the couch. He opens the drawstring and starts groping for his phone. Mel returns her attention to hers.

Shree descends the ladder, hoping to see if Pando wants any company. "Morning, sleepyhead," she says to him.

He looks up from the backpack and sees Shree's shining eyes. "Hey, Shree," he says with a smile. "Sup."

"You are, finally," she replies. "You've been asleep all day!"

Pando lets out a short laugh. "Yeah, I was fucking tired."

"Apparently," Shree says. "Can I get you anything?"

He shakes his head. He resumes fiddling with his pack, pulling out his phone. It isn't a slate like everyone else's, but a rather minimal flip phone. He powers it on.

The microwave beeps. Mel pulls the door open, revealing a cloud of smoke and the foul odor of burnt cabbage. The venomous scent hits her nostrils like a sledgehammer.

"Shit!" she cries. "The fuck did I do wrong?"

She grabs the paper box and rereads the instructions. "Microwave on low power for 3 minutes," it reads.

"Oh, fuck you, 'low power,'" she says to the box, throwing it in the trash along with the rest of her spoiled food. Life is so annoying. She heads for the cabinet instead.

"What happened?" asks Shree.

"Nothing, just ruined my food. It's whatever, I didn't want to eat Salisbury steak anyway." In replacement Mel has pulled out a sleeve of noodles, and she grabs a bowl to put them in. She fills it with water, putting it into the still-open microwave.

"Alright, let's actually read it this time," she says to herself. Two minutes on high. She slams the door closed, pushes a button twice, and like magic, her new food is being heated again.

Pando rummages through his bag again. He pulls out a yellow plastic bag from the bottom of the sack. He takes a seat, dropping his backpack between his feet. Shree takes a seat beside him.

From this yellow bag, he pulls out a lighter, a silver grinder, a multi-colored pipe with a chamber for water, and a small, clear bag. It's filled with pungent green plant flowers. He lifts it up in the air for his hosts to see.

"You guys don't mind, do you?" he asks.

Mel glances over to the stranger on the couch. "Only if you share," she remarks.

"Of course," he says. "What are you making?"

"Ramen."

"Oh man, ramen would hit the spot right about now," he says.

"There's more in the cabinet if you want one," says Mel, motioning to the relevant door. "Chicken flavor."

Pando rises to his feet ecstatically, tossing the bag of drugs onto the coffee table; he holds onto the pipe. "Don't mind if I do," he says happily. "You want some ramen, Shree?"

She agrees, prompting Pando to join Mel in the kitchen. The microwave beeps again, and Mel grabs her steaming bowl, yielding the space to the newcomer. She joins Shree in the living room.

Cooper, studying in his room, hears the commotion below and decides to go down too.

"Hey everybody," he says, taking a seat beside Mel.

Pando puts two bowls of ramen in the microwave. He turns the faucet on, filling the bubbler pipe with a supply of freshwater.

"Oh hey, Cooper," he says. "It's nice to see you again."

Cooper nods with a smile. "Hey Pando, how you doing?"

"Better, now that I've slept," he says. "You guys have a really comfy couch." He sits again next to Shree, leaving the microwave to work its power. He grabs the bag of weed off the table, opening it tenderly and pulling out a small nugget. He lifts the top off his grinder, inserting the plant between its pointed teeth. He closes the contraption and gives it a solid crank. He lifts the lid to reveal the ground up herb, which he spills into his palm. With this, he packs a bowl.

Cooper eyes him uneasily during this process. This guy has made himself at home fast.

Pando flicks the lighter on and takes a hit from the pipe. After he exhales, he hands it to Mel, who puts her soup in her lap and lights up. She hands it back to Pando.

"Shree?" he asks, offering her the bubbler.

"Yeah, okay," she says, grabbing the pipe and lighter out of his hands.

"What about you, boy?" Pando asks.

Cooper thinks for a second. "I've never smoked before," he says cautiously. "But my coworkers do all the time."

"It's never too late to start," Pando says, handing Cooper the bubbler. "Here ya go, man."

Cooper accepts without much forethought. After all, he'll have to try marijuana eventually, right? Besides, literally everyone else he knows in this city smokes, including, apparently, Shree.

He brings the bubbler to his lips, which has been shared freely with the germs of who knows how many people. Igniting the flame over the green crumbs in the bowl, he draws in a breath, pulling the heated THC down into his lungs. He coughs out smoke wildly.

"First hit is always the worst," says Pando.

Cooper continues to hack up a lung. He regains himself after a brief struggle with irritation. He's had his first hit of weed, and he survived.

Mel slurps her soup loudly. She turns to Pando. "So what's your story?"

Pando exhales a large plume of exhaust. "I'm a climate refugee," he says sullenly.

"What's that?" Mel asks.

Pando hands her the bubbler. "You know Typhoon Powell?"

"Nope," she says, grabbing the pipe and lighter.

"Well it destroyed my mother's village, and killed my brother," he goes on. "I just happened to survive."

Mel breathes out smoke and passes the pipe to Shree. "That sucks, I'm so sorry," she says.

Pando raises his eyebrows and nods. "My mom's still out there, cleaning up the mess. And in the meantime, I'm on my own. Hence why I'm couch surfing with you guys right now."

The microwave blares that it's finished; Pando gets up giddily. He pulls out the bowls and returns to the living room, setting one down in front of Shree.

"So wait a minute," cries Mel. "When did this storm happen?"

"Three weeks ago?" says Pando with uncertainty. He hasn't exactly been counting how much time has passed since his world came to an end.

"How come I haven't heard about this?"

Pando shrugs. "I dunno. Do you read the news?"

"Nope."

Pando grabs the bubbler from Cooper's hand and tears a thick flame from the lighter. He takes a gigantic hit, finishing the pipe's supply. He reaches for his bag of weed and begins grinding more.

"I bet you don't know about climate change either," he adds as he spins the cold metal. He empties the bud onto the table and starts repacking the bubbler.

"Not even a little bit," says Mel.

Pando shakes his head in disbelief. "What about you two?" he asks, his attention on Cooper and Shree. They shake their heads no.

"Of course not," he says, lighting the bowl. He draws another huge hit of smoke into his lungs, and blows it into the air almost furiously. He passes it to Mel.

"So if I were to tell you guys," he continues, "that by the time we're middle-aged, the coldest years in this city are gonna be hotter than the hottest years today, you wouldn't have a clue what I'm talking about, am I right?"

Mel blows tree; Cooper and Shree look at him silently.

"You guys don't know anything about this?" he asks, clearly frustrated. "About how by the end of the century, half of this city will be underwater? About how the West will be in permanent drought? About how the food chain will have completely collapsed?" His silver eyes stare right through them, each question more demanding than the last. "You guys know nothing about this shit, do you?"

Mel coughs out a lungful of smoke, waving her hand through the cloud to dissipate it. "Oh please," she says dismissively. "Climate change is a load of crap."

Pando throws her an enraged look. "As someone whose life has just been destroyed by climate change, I'm gonna pretend you didn't say that," he gibes.

She relents, ignoring the antipathy in Pando's comment so as not to unduly start an argument with their new guest. Nonetheless she believes, as many people do, that the science on climate change is far from settled. Pando, she figures, is simply a victim of bad luck.

"Hold on a sec," says Cooper. His mind is racing wildly down hazy avenues of thought. He's not sure if it's due to the drug or the things he's hearing. "So you're saying that climate change is gonna flood New York?"

Pando turns his attention to Cooper. "Dude, forget that. Climate change is gonna drive a good portion of life on Earth to extinction."

Cooper's face flushes at the thought. Extinction? "Jesus," is all he utters.

"That's just the tip of the iceberg," Pando says. "Every seven minutes, a species gets wiped off the globe forever. That's what's happening out there. The world is literally falling to pieces around us, and none of you know a thing about it."

Cooper blinks, feeling a naivety like nothing he's ever known. A mass extinction — global death. Food chains unraveling like spools of yarn. Humans may be perched comfortably at the top of that chain, but he knows it's only because we depend on every link in the chain under us, from bacteria on up. If those links come undone, it's only a matter of time before we do too.

"That's nonsense," Mel disagrees. She privately labels Pando as just another radical environmentalist, the kind of person who's been brainwashed by ideology into believing threats that have no merit. To her, this guy's a sham.

Shree sits in silence, listening intently to what Pando is saying. She hasn't yet taken her turn to smoke from the pipe she's been holding for far too long for Mel's taste.

"Are you gonna hit that or what?" she asks her.

Shree looks at her blankly, still digesting Pando's words. She shakes her head and puts the bubbler on the table; Mel picks it up.

"Here, Cooper," she says, handing him the pipe.

"What do you mean that's nonsense?" Pando asks pointedly.

"There are always people like you that say the world is gonna end," Mel explains. "Bacteria have probably been saying it to themselves since the dawn of time. Well guess what, we're still here."

He laughs sharply. "The difference is, this time it's actually true."

"Hurricanes happen all the time, dude," Mel says with annoyance.

Pando glares at her angrily. "Y'all need to start paying attention," he condemns.

Cooper considers this charge, remembering his conversations with Natasha. As an Ivy Leaguer, he regards himself as a rather well-educated person. But his higher education never mentioned anything like this. Never mentioned the world he's preparing himself for won't resemble the one he'll actually live in.

"Maybe it's just too hard to explain," he suggests. "Like maybe climate change is just too complicated for most people to understand."

"What do you mean?" snaps Pando. "I *just* explained it to you. The climate's changing. Everything is dying. It's as simple as that."

Cooper bites his lip uncomfortably. Everything is dying. Does that mean he's dying too? If the world dies, he dies with it. It actually is as simple as that.

"So what do we do?" he asks desperately.

Pando reclines in his seat cushion. It's not an easy question. "Well, it's getting hotter because of greenhouse gases. You've heard of those, right?"

"Uh, yeah," Cooper growls. "I wasn't born yesterday."

"Okay, then you know what I mean," Pando offers. "When we burn carbon, it gets hotter."

"Like me right now," says Mel, lighting up a hit of weed and blowing it into the air. "Boom. Greenhouse gas."

Pando rolls his eyes. "Yeah, sorta. But carbon stays in the atmosphere for thousands of years. So once we burn it, it stays there. We can't undo that."

The boy's eyes go wide. *Thousands of years.* His brain can't even register the scale. He can barely see the present, the enormity of his world and his microscopic position in it. What it really means to be one of *billions* of humans. How much they eat. How much they drink. How much carbon they burn.

"That's fucking insane," says Cooper. *Thousands of years.* That would mean carbon burned in the belly of the *Titanic* is still in the air today, warming the world over a century on. One catastrophe feeding right into another.

"It is insane," Pando agrees. "Even if we stop burning carbon today, the planet's gonna warm for a long, long time. Whenever you think it's getting hot out, just keep in mind that this is only the beginning."

Cooper swallows hard, remembering the recent heat wave in October. Never before had New York City trick-or-treated on an 80-degree day. He and Mel carved pumpkins in shorts with the fan on. He didn't think much of it at the time, but now it makes a lot more sense.

Shree is still silent, staring down at the table in contemplation. As a teacher, she's been taught about climate change as well as anyone who's certified to teach. She passed the tests, at least. But never has she really considered what it actually means for the people she cares about the most: her students, and through them the future.

Mel, growing tired of this avenue of conversation, finishes her soup and stands from the futon with bowl in hand. "Well, thanks for the positivity," she jeers. "I'm gonna go upstairs. Pando, it was nice to meet you."

Pando watches her rinse her dish in the sink before climbing to her room. He feels like he's started off on the wrong foot with her — but seriously, she should know better than to question climate change in front of him like that.

He looks over to Cooper, whose puckered forehead indicates he's ruminating on what he's just heard. Shree, meanwhile, looks completely distraught.

All three millennials are blasted, and Pando's new roommates have become morosely quiet. A gloom has settled over the room, coloring their thoughts a dark shade. His words have left them stunned, an unshakeable dread in the pits of their souls.

"Anyway," he says to alleviate the silence. "That's my story."

A notification buzzes from Cooper's phone; he pulls it out to check the text.

Mel: r u still listening to that quack?? 11:51pm

Shree appears to have received the same text, for she is now checking her phone as well.

Cooper ignores it, powering off his screen. Curious as to the time, however, he hits the side button again, the display coming back to life. It's almost midnight. Time flies.

"Jesus, it's later than I thought," he says. "I think I'm gonna go to bed too."

He rises slowly, eyeing Shree across from him. She looks like she's about to cry, or at the very least could use a hug. He figures he'll leave that to Pando. He suddenly remembers he still hasn't eaten dinner, not that he had the money for food tonight anyway. He doesn't mind, however; the nausea in his stomach has quite taken care of any appetite he would have had. He climbs the rungs of the ladder haltingly, glancing back at his roommate, and at the refugee that has shaken up his world.

"Goodnight guys," he says softly.

Shree shifts her gaze to the boy climbing the ladder. He's so thin, so bony, vulnerable. She pities him as his frame shudders up the steps.

"Goodnight, Coop."

He silently closes the door to his loft. Shree and Pando are left alone downstairs.

"Well, I think you made a great first impression," she says to the refugee.

He chuckles somberly. "Sorry, I get carried away sometimes," he offers in his defense. "I like to speak my mind."

"No, don't be sorry," she reassures. "I agree with you. You were very...honest."

Pando smiles discouragingly. "I tell it like it is. The truth hurts. I don't have time for bullshit."

"I wish more people were as honest as you," Shree remarks. "A lot of the people I've met lately seem to be a bunch of liars. Honesty is a precious commodity these days." Like a hipster fawning over an audiophile LP, she values honesty above almost all other character traits. And Pando has certainly left an impression.

"Most people only lie because they feel like they have to," he speculates.

"Maybe," she considers. "All the guys I've met through online dating were never really honest. They all come on like they're looking for a relationship, but before I know it they're just talking about sex and asking me for nudes."

"What's wrong with sex?" he inquires. "Everybody likes a good hookup now and then."

She chuckles. "I can't disagree with you there," she says, smirking at him. "But like, if that's what you're interested in, then just be up front about it. Don't beat around the bush and pretend you're looking for a relationship when all you want is to get laid, you know? Does that make sense?"

"Yeah, but I think if guys were up front about it then you wouldn't even talk to them," Pando wonders.

Shree sighs. "Yeah, that's true. I mean I don't know. I want to get laid too."

Pando laughs. "Good to know. Well I'm right here in case you need me." He smiles at her.

She smiles back and squints into his eyes, skeptical at the invitation he just came forward with. "Your eyes are super red," she tells him. "You're baked."

"Hell yeah I am!" he shouts.

"Me too," she sighs, looking around at the otherwise empty room. "So."

He looks at her curiously, waiting for the rest of her comment.

She looks at him suggestively. "What do you want to do?"

Pando puts his hand on his chin, thinking honestly about her question. After a hot sec, he arrives at an appropriate answer. "You want to watch SpongeBob?" he asks genuinely.

"I thought you'd never ask," comes her immediate reply. With all the talk of extinction fouling her mood, a generous dose of harmless children's comedy feels the perfect cure for her restless nerves. "I'll go get my computer."

She leaps off the couch like a spring, only for the energy to drain from her body as soon as she reaches her feet. Her vision goes purple; a rush of blood to her head. She falls back onto the couch.

Pando looks at her with concern. "You okay?"

"Got up too fast," she explains, rubbing her palm against her forehead. She gets up again, this time with less vigor.

Pando watches her admiringly as she enters her room. He puts his hands behind his head as he leans back into the couch cushion, appreciating the simpleness of the room: its utilitarian layout, its intimate common area in the middle, its bare concrete floors.

I could get used to this, he thinks, smiling to himself in his high stupor. If nothing else, at least he has this.

Shree comes back down the ladder with her lightweight computer. She takes a seat next to Pando on the orange couch, placing the laptop on the table in the center. Her hands fly across the keyboard, and with a few quick commands, she pulls up an episode of the legendary turn-of-the-millennium cartoon that defined their childhood. She clicks a button, expanding the video to full screen. "*Are you ready, kids?*" comes the familiar question.

"Aye, aye, captain!" Pando calls in response. He can't even guess how many times he's heard these opening credits.

Shree giggles. "This is just what I needed," she says. "Let's forget about the bullshit of adulthood and be kids again."

"Amen to that," he says.

Beds are Burning

The boy flails about in subfreezing green-blue water. Pain is the dominant signal in his mind; the terror, in this moment, has been relegated to a distant abstraction. Gasping for breath, he looks around in a panic. It feels like he's been here before.

He isn't alone in the paralyzing sea. Hundreds of other passengers are treading water with him in the salted ice water. Some of them bob gently, floatation devices keeping them at the surface; the rest of them either clutch odd bits of flotsam or merely struggle to remain above the waves. Cooper knows it's a waste of effort regardless; now that they're off the ship, they're as good as dead.

He looks up at the rearing liner from which he just jumped. Its enormous black sides are beginning to fold outward, straining under the huge weight of the stern, now hoisted far into the night sky. It rises perilously, the giant propellers climbing farther and farther upward. These now dormant engines once carried him powerfully through the waves, channeling the tremendous supply of energy liberated from coal-fed boilers. Like most of the passengers, Cooper was looking forward to his new life in New York, a life of prosperity and opportunity. But all their ambitions have been savagely curtailed by the deadly iceberg which lay obstinately in their path.

Even in its death throws, the ship is a marvel to behold. Man's ultimate statement of progress. An unsinkable ship. Yet here she founders, the death toll innumerable. When the danger was spotted, the crew doubtlessly reacted with all due urgency, but it was far too late; their fate had already been sealed. The propellers were too slow to react, unable to overcome the sheer momentum of the massive vessel's forward movement. Instead of carrying them safely to their destination, the engines ultimately steered them straight to their doom, a victim of the fundamental limit: nature always wins.

Thus Cooper drowns in the Atlantic, watching an unbelievable catastrophe from the knowing vantage of one of its casualties. He thought that by jumping off the edge of the deck as he did, he would be hastening his own demise; he didn't expect to survive a hundred-foot dive. Instead, he landed in the water alive, bruised but still breathing, still suffering. With nothing left to do, he sloshes through the frozen seawater, regretting ever setting forth on this doomsday voyage.

Metal groans overhead; he's directly underneath the colossal end of the sinking vessel. With an ear-splitting shriek, the thick sides buckle outwards, splitting the impregnable hull in two. With the back of the ship broken, the considerable weight of the stern succumbs to gravity, severing itself from the compromised bow. It plunges downward, spraying up a tidal wave of white saltwater. Cooper watches from below as he lets the iron giant overcome his tiny body.

His eyes open with a sharp breath. He's strangling his pillow with the force of a vice grip, panting uncontrollably in his bed.

Just a dream, he assures himself. He takes several deep inhalations, hoping to calm his racing heart. He grabs his phone and flicks on the screen. 6:20am. He lays back down, waiting for sleep to take him again.

Shree wakes up next to Pando on the couch. His arms are wrapped around her loosely; he's fast asleep in his dirty clothes. She gently pushes herself up from the couch without disturbing the sleeping man next to her. Her grogginess limits her thoughts to the basic demands of homeostasis. She walks mindlessly to the kitchen, grabbing a plastic cup from the faux-wood cabinet. She flips on the faucet, catching its life-bringing water into the clear receptacle. She chugs it zealously, every parched cell in her body rejoicing in the cool drink flowing down her throat.

Her eyes catch the clock illuminated on the microwave display — 8:48am.

Oh shit, I'm gonna be late for church, she thinks. She finishes off the glass and grabs a bagel out of the fridge. She pops it into the toaster and makes for her room, the bagel browning in solitude.

She forages her closet for a nice outfit, selecting a soft greige blouse. She changes quickly, tossing yesterday's clothes into her laundry bin. She grabs her keys off the hook and pulls the door closed behind her.

Her bagel is done. She pops it out of the toaster and slathers it in strawberry cream cheese. With breakfast in hand, she trots out the door.

The late November air is refreshingly cool on her face. She's only wearing a light jacket, but with her body burning calories as she walks, she's more than content with her choice of attire. People tread past her on the sidewalk: young people in flat-brimmed hats and distressed jeans; fit people in sporty athleisure wear; old people in button downs and pressed pants; families with boisterous children in tiny sneakers, headphones over their ears. It's a lovely morning in New York City.

The church is only a few blocks away. Its white spire pierces skyward, an arm extended as high as its reach allows, striving to touch the face of God. Shree walks through the double doors into the holy building, hanging her jacket on the rack by the entrance. The sermon has already begun.

She enters the nave quietly, taking a seat at the first available pew. She is one of the younger members of the congregation this morning; aside from a mild-looking young man sitting with an elderly couple, she's the only 20-something in the room.

The pastor delivers his message from the pulpit at the front. A cross features prominently behind him, with rows of candles lit serenely at its feet casting a flickering glow. From its sides hangs the crumpled body of Jesus, his despairing eyes staring lazily at the floor, weary from the sins of humanity. The bloody crown of thorns juts from his long, tattered hair. The Son of God must have welcomed death by the time it finally claimed him.

"Jesus loved us all," the pastor bellows. "God gave his only Son for our salvation, and we repaid his gift by sentencing him to death."

Shree isn't paying much attention to the words reverberating off the chamber walls. She's transfixed on the bleeding man depicted in the center of the room.

"He died for the redemption of our sins," the pastor continues, "yet we continue to transgress against all of creation."

All of creation, her thoughts echo. Her attention snaps to the pastor, and she heeds carefully the words that follow.

"Does Ezekiel not teach us this lesson soundly? Relating the demands of our holy God, he says: 'Woe to the shepherds of Israel who only take care of themselves!'" The pastor raises his voice in condemnation. "'Should not shepherds tend the flock? You eat the curds, clothe yourselves with the wool and slaughter the choice animals, but you do not take care of the flock.'"

The pastor's sermon plays in her head long after the mass, echoing through her thoughts into the early afternoon, repeating select phrases over and over again as she ascends the stairs to her apartment. She's trying to remember every word, eager to share her revelation with Pando. She understands now; she understands what he means.

She finds Pando lying on the couch listlessly, sleeping without a sound. Discouraged, she grabs her computer off the table and hikes up to the loft. She looks back at the refugee below. The man's brother is dead; we didn't tend the flock.

In the adjacent room, Cooper has been on his laptop for hours, researching climate change the entire morning. He's jumped from article to harrowing article, trying to wrap his mind about this complicated evil that now seems to threaten everything he's ever known. The implications range from the bizarre to the downright terrifying, a world more akin to science fiction than anything that exists today. In his haphazard research, he has amassed over two dozen webpage tabs along the top of his browser window, each one describing some seemingly unreal element of climate catastrophe, each one a true story.

"Wildfires continue to burn outside LA as drought enters fifth year; fire season now year-round," reads one headline. The thought of endless forest fires instills in Cooper images of Doomsday: a perpetual battle against the flames, engulfing one community after another. Entire neighborhoods evacuated, thousands of people homeless.

Another webpage entitled "Ocean acidification on track for worst-case scenario" illustrates the breadth of the issue: not only is the land burning, but so are the seas. Life everywhere appears to be perishing on an unprecedented scale, at a rate humanity has never before encountered, accelerating beyond all our wildest projections.

He doesn't know what to make of it. He just knows his hands are clammy, and he's still vaguely nauseous. Nevertheless, his body reels from malnutrition; it feels like a saw is tearing his stomach in two. Obeying the commands of biology, he works himself up from his bed, in search of forage or prey.

Pando rises from his slumber later that afternoon, the sound of hip hop music from downstairs announcing this fact. He reclines on the couch, nodding his head to the beat as he operates the computer sitting on his chest. Through the wonder of prosumer technology, he's editing a rhythm-heavy backing track on a music app. This loop, he hopes, will be the genesis of a groundbreaking statement, the first step in a lyrical creation over which he intends to lay a rhyme.

He sees Cooper exit his room and make the descent along the ladder.

"Sup Pando," Cooper greets.

"Hey man," he says, too absorbed in his music to pay much mind to the kid.

"Whatcha doing?" Cooper asks, eager to start a conversation. He secretly hopes he'll put the computer down so he can have his full attention.

"Just chillin', dude," Pando drawls.

The air is thick with the scent of marijuana, another telltale indication that the couchsurfer had awoken from sleep. The bubbler sitting on the table holds a container of spent ash. Cooper takes a seat on the pink futon.

Pando doesn't stir, scarcely acknowledging Cooper's presence. He knows he's just a guest in the boy's apartment, but he's trying to communicate that he's kinda in the middle of something, without saying so in words.

Cooper reads this, but he disregards it anyway. He's been itching to speak with the refugee all day, still on edge from the conversation last night. The discoveries he's made on the internet all day haven't served to ease his turbulence.

"Can I talk to you for a minute?" Cooper asks gingerly.

Pando shifts his attention from the glowing screen. He raises an eyebrow at the small white boy sitting across from him, studying him carefully. He notes his messy hair, his bony face, his thin arms. He looks like a skeleton, wasting away. So young, so vulnerable. So helpless, like his brother back in Kabubwason.

He sighs mentally. He lifts his computer onto the table and sits upright. "Sure, Coop. What's up?"

Cooper clasps his hands uneasily, balancing his elbows on his knees. He's staring at the floor, shaking his head uncertainly.

"I'm just..." he trails off, thinking about the tragedy unfolding in the Philippines. Despair wells up within him uncontrollably. People are dying. *Everything* is dying. And he doesn't understand. He looks up to the ceiling, confused as to what he's feeling.

"I don't really know, actually."

Pando scrutinizes him closely, trying to read his thoughts. He's waiting for the kid to speak up.

Cooper knows what he wants to say, but he hesitates; maybe he doesn't actually want to know the answer. He looks over to Pando, whose dirty white bandage is peeling at the corner. He's waiting.

You have his attention, Cooper thinks. *Just ask him.*

"I'm listening," Pando says — not impatiently, but to remind Cooper that he doesn't have to hold back.

"Is everything you said true?" he blurts at last.

Pando blinks. "What, about my hometown being destroyed?" he asks. "Of course it's true."

"No," Cooper says. "I mean about global warming."

Pando's face softens. "What do you think?" he asks rhetorically.

"I think you've been exaggerating," Cooper lies. Between Natasha and the internet, he's had his fair degree of exposure to the topic, but he still doesn't want to believe it's true. How can everything be falling apart so completely? How can it be?

"What, you want me to cite every article I've ever read?" Pando snaps. "Cooper, my home was wiped off the map by the most powerful hurricane to ever wash over land in *recorded history*. My brother is dead because of it."

Cooper nods, breaking eye contact to stare at his feet.

"And the reason it was so strong is because the Pacific Ocean is hotter than it's ever been. And why do you think that is?"

"Global warming," Cooper answers.

"It isn't hard to connect the dots," says Pando. "Believe me, I've been reading about this for years. That hurricane was only a taste of what's to come."

If it was acceptable to cry at this point, Cooper would do it soundly. But he's an adult, or trying to be one. Adults deal with devastating news with poise and control. Instead of revealing the emotions brewing underneath his tenuous appearance of calm, he lets his mouth ask the questions on his mind.

"How bad is it going to get?" he asks, voice cracking.

"So bad that—" Pando pauses. Maybe it's best not to start with that. "Alright, let me put it like this. We burn fossil fuel, right?"

Cooper nods, glancing at the scene around him; everything in the room has been touched or influenced by oil. "And the more we burn, the hotter it gets," he says.

"Right, good." They're on the same page so far. "The thing is, carbon only part of the problem."

Shree has been in her room this whole time, grading her students' homework assignments. She can't help but overhear the conversation playing out below through her paper-thin walls. She rises from bed and opens the door a crack to listen more intently.

Pando continues explaining climate change to Cooper. "The Earth has a number of what are called positive feedback loops," he describes.

"What are those?" Cooper asks curiously.

"Basically they're the reason we're super fucked," Pando says candidly. "I'll give you an example. The Arctic sea ice is melting, right?"

Cooper nods silently.

"Well ice is white. White things reflect energy, as opposed to dark things, which absorb energy. That's why you don't go outside wearing a black shirt on a hot summer day, because your shirt is just gonna soak up sunlight, and you're gonna be really hot."

"Right," says Cooper.

"So as the world gets hotter, instead of having a bunch of reflective ice floating around, we have dark blue ocean," he says. "The ocean absorbs heat, which melts more ice, which exposes more ocean." Pando spins his hands in a rolling motion, indicating a perpetual cycle of increasing, runaway heating.

"I see what you mean," says Cooper, registering his explanation.

"Right, well that's a feedback loop," says Pando. "There are a lot of them. Another one involves permafrost."

"What's permafrost?" asks the pupil.

"It's frozen soil," Pando explains. "Like the ice sheets, the permafrost is thawing all over the world. But permafrost contains a lot of methane."

"Methane? Like the stuff farts are made of?" Cooper asks coyly.

"Yes, that methane," says Pando. He would smile if it wasn't so deadly serious. "Methane is a greenhouse gas like carbon dioxide, except it's twenty times more powerful."

"Shit."

"Exactly." He's glad the kid is beginning to understand. "When permafrost thaws, it releases that methane, which warms the planet even more, which thaws more permafrost, which releases more methane. I think you get the idea."

Cooper agrees solemnly. "Basically, we're screwed," he says.

Pando nods in agreement. "That's why we have to stop burning carbon, like, *yesterday,* because after a certain point, these feedbacks are gonna compound on each other, and then everything is gonna go to hell. We won't be able to stop them. And when that happens, it's all over. Planet Earth is fucked."

Planet Earth is fucked, thinks Cooper. "When does that happen?" he asks.

"It's probably already happened," Pando says bitterly.

Shree, hearing this, decides to join the guys downstairs and listen to Pando describe the end of the world. He notes her arrival as she leaves her loft, getting the sense that she's been listening the whole time.

"It has?" asks Cooper.

Pando shrugs as Shree silently takes a seat next to Cooper. "Probably. I mean there's no way to know for sure," he admits. "The politicians are trying to keep global warming to two degrees Celsius, but scientists think the feedbacks are gonna kick in way before that." The man shakes his head. "Honestly, any degree of warming is playing with fire, but at this point it's anyone's guess when we'll cross the point of no return, providing we haven't already done so."

Shree doesn't take her eyes off him. His silver eyes radiate the intensity of twin suns, yet his soul feels shrouded in darkness. He sounds like a corpse, retelling the story of his own death. The acrid experience of Typhoon Powell seems to have erased any hope he had for the world, and all that remains is the dead, strangled shadow of his former life.

"And how much global warming have we already caused?" asks Cooper.

"About a degree," Pando answers factually.

Cooper feels a cooling rush of relief wash over his body. "Oh, well that isn't so bad," he says. "That means we're only halfway there."

Pando shakes his head fervently. "Not exactly."

The boy's muscles tense again. His chest constricts.

"Every day we burn more carbon than we did yesterday," says Pando. "And like I said, it stays there for thousands of years."

Cooper gulps, embarrassed to have thought things weren't as bad as he feared. With this reminder, the future still looks as bleak and desperate as ever. The boy's heart sinks; it's already too late.

A door creaks open from the loft. It's Mel; she's heading downstairs. Headphones blast music into her ears, allowing her to ignore any conversations taking place in her absence. She traverses the ladder without a sound, slipping quietly into the kitchen. Nobody pays her any mind.

A wild idea comes into Shree's head. "What if we just move north?" she asks innocuously.

Pando looks at her curiously. "We can't just hide from global warming. You know how hard it is to move a city once it's already built?" Shree shifts her eyes to the floor.

"Our cities will be underwater soon anyway," he goes on. "Once the polar ice sheets begin to collapse, the coastal cities will flood over."

"Do you know which ones?" asks Cooper.

"Most of them," he answers.

"Miami?"

Pando chuckles. "Miami? Miami will be underwater in just a few decades."

An outpouring of dread inundates Cooper's heart. *Allie. Dad.*

"Over half the human population lives in cities," Pando says. "Imagine the throngs of refugees like me. All hell will break loose. The fabric of society will fall apart."

"Not if we go underground," says Shree, searching for a solution.

"Going underground isn't going to prevent starvation," Pando counters. "The food chain will unravel as species go extinct because they can't adapt to their changing environment. Humans won't be the only refugees when all is said and done."

Mass extinction, Shree thinks menacingly.

"We can't eat if there isn't anything *to* eat," he continues. "Our crops will wither in the extreme heat. Our farmland will dry up as the land desertifies. And that's before we even think about the ocean."

"You mean acidification?" Cooper winces.

"Exactly," says Pando, surprised the boy has even heard of the term. "The oceans absorb a large amount of carbon from the atmosphere. It's already getting to the point where some creatures are beginning to dissolve from extreme acidity."

Dissolve from extreme acidity. This really is the end of the world.

"Fuck," Cooper utters.

"Yeah, it's a shit show out there," says Pando, pointing out the obvious.

Mel is rummaging around in the kitchen, making herself a tuna fish sandwich. The volume from her music drowns out all other sound coming into her ears; she can't hear Pando, and wouldn't want to anyway. She hikes up the ladder with her lopsided fish sandwich, almost dropping it as she makes the ascent. She withdraws back into her room, closing the door behind her with the click of a lock.

Cooper and Shree both feel supremely defeated, like their lives suddenly hold no value now that they know they're just another pair of humans contributing to the inevitable collapse of the world. Cooper puts his face in his hands, shaking his head defiantly. Everything Natasha has said is true, it seems. He pulls his hands through his hair as he leans back against the couch.

"Now wait a minute," he says accusingly. He eyes the refugee with contempt: his clothes are filthy, his face wounded. But he's still alive. And as long as he's alive, he should be doing something, right?

"If you know so much about this, then what are you, like, gonna *do* about it?" he asks desperately.

Pando leans back unhappily and shrugs. "I dunno, dude," he sighs. "What are you gonna do about it?"

The boy just stares.

"I dunno if there's anything *to* do about it," admits Pando. "I'm 24 years old. Other people run my life. As long as I operate in the system, I'm guilty. And so are you. And there's nothing we can do except wish it didn't have to be this way."

Cooper's gaze falls, recognizing his point. Though armed with foresight of their own demise, none of them are in a position to do much of anything about the widespread collapse of human civilization.

The words sit with Shree with the weight of the world; it makes perfect sense to her. She's part of the problem, just like everyone else. And to her, as to Cooper, the idea of contributing to the sins of humanity is almost too much to bear.

"If I'm being honest," says Pando, "I don't plan on sticking around for what's coming. I'll wander from job to job for a bit, earn my fifty grand a year, eat food, burn carbon, and probably kill myself before it all falls apart."

All the eyes in the room look over to him like he's crazy. *Suicide?* Only crazy people think about that.

"You can't do that, man," says Cooper. Suddenly this feels like a counseling session.

"Why not? Aren't we all killing ourselves?" Pando asks. "We're doing it all the time, with every minute we burn. Sometimes all you can do is look at the world and decide not to be part of it."

Nobody speaks. Nobody looks up. Nobody wants to live in a suicidal world.

Shree shifts her body agonizingly in her seat. *Suicide is a sin,* she thinks to herself. *We can't all be suicidal, can we?* She listens for an internal reply, her voice of reason to assure her worrying mind that everything is going to be okay. The voice never comes.

"I dunno about that," says Cooper.

"Think whatever you want, Coop, but we're still on track for four degrees Celsius by the end of the century," Pando reminds him. "And that's suicide."

Cooper feels terribly dissatisfied by all this talk. It's too speculative, overblown, abstract. He needs something more concrete before he's willing to allow himself to believe Pando's words. Typhoon Powell may still be a tragic fluke, the torrential downpours mere coincidence. If he's going to give in to the idea of climate change, he has to find more evidence first. It isn't enough to be 95% certain. It has to be a hundred.

"I think I'm gonna take to the interwebs," he announces, standing up from the couch. "No offense, Pando, but you still kinda sound like a crazy person."

Pando snickers quietly. "So do you," he says, waving him up the ladder. "Go prove me wrong."

Shree is trapped by an incapacitating thought. She looks like a statue to Pando, her muscles stiff, a frozen expression of sorrow chiseled into her face.

"What's wrong?" he asks her, fully expecting the answer to be "*Everything.*"

"Nothing," she breathes. "Just thinking about my kids."

"It's gonna be rough for them," he replies with a grimace. She looks unhappy with his answer. "I'm sorry if I make everything sound really negative," he adds.

"You're just being honest," she reasons. She smiles weakly, but her attention is drawn down in her lap. Suddenly, and without inhibition, she squeezes a tear out of her eye. It falls onto the cement floor below.

Pando slides over to her and puts his arm around her shoulders. "Don't dwell on it. It's only gonna make you sad."

She sniffs. The advice is sound, if impossible to follow.

He puts his finger under her chin and lifts it up tenderly. "Hey," he says, looking into her dark eyes with his penetrating silver irises. "You wanna smoke?"

She closes her eyes and smiles. He really does care.

"Come with me," she says, rising to her feet. He gets up after her, snagging his green medication off the table. She grabs him by the hand, leading him silently up to her room. She closes the door with a thump. There's a giant stack of papers next to her desk.

"You see all this?" she cries, gesturing to the pile. "I have to grade all this shit."

"Ew. Fuck that."

She sighs. "Where's that bubbler of yours."

They take a seat on the bed while Pando prepares the drugs from his bag. He pulls out a lighter and they share an easy smoke.

"I feel like this stupid curriculum has me testing these kids every other day," she moans. "They're all so stressed, and me too because of it. This teaching thing is nothing like I thought it would be."

Before she hands him the pipe, she pauses to examine him curiously. His sterling eyes peer into her, stars in the nighttime sky. His face is tense, like he's always on alert, but his skin is soft, his demeanor tempered. He may be the victim of forces beyond his control, but in this moment, he allows himself to appreciate the tender moments that still make life worth living, at least for now.

She slowly raises her hand to touch the bandage over his cheek, bringing her finger to its edge to fiddle with the loosened adhesive. Seeing no signs of protest, she continues to peel it gently back, down to his chin and off completely. Amidst the moist flesh of his jaw, embedded into its pockmarked surface, stares a deep, jagged scar, two inches long and as thick as a pencil. The stitches are spaced haphazardly along its sides; he may have sewn it himself.

Pando smiles.

. . .

It's 11:30am. The bed is empty beside him, a ruffled fold of sheets. Shree has long since left for another arduous day of teaching. It's Monday morning, and Pando has no plans.

With only one window by the sofas to provide natural light, the living room is generally dark without the incandescent. But Pando has arrived at the rare time of day when the sun is shining through the window in a narrow beam, bathing the chamber in amber light as it reflects off the white walls. It feels like a sanctuary of gold, a holy place where he can have communion with his remaining treasures in life: his computer and his weed.

He pulls out his laptop from under the couch and flips it open. The screen loads to a music streaming app, tiles of album covers filling the page. He selects an option titled "Ultimate Chill-Out Playlist," determining it to be a good complement to his activities this morning.

The sack of marijuana on the table is still open; he picks out a nug and grinds it finely. Loading it into the bubbler, he inhales a lungful of gaseous carbon and vaporized cannabinoids, sending his mind off to sensory bliss. He takes another hit, then another. In short time, the bowl is kicked.

The music rolls smoothly out of the computer speakers, carrying him along drug-induced highways of his mind. The world becomes the music to him, and he finds himself unable to escape its groove. It has unfettered command of his thoughts, his body, his soul.

He snaps to, remembering he's still waiting for an update from his family in Kabubwason. He opens another tab on his browser and navigates to his email page. It loads in an instant, displaying the list of messages that have piled up over the life of his digital existence: 6,718 unread emails, the heading of each displayed in bold font. Seven of them are from this morning alone: an ad for a skin care product, local job openings, and a bank account summary, to remind him how poor he is. Not one of them is from his mother.

His heart fills with disappointment. He hasn't seen or heard from her since that fateful day. Maybe she meant it this time; maybe he truly is all on his own now.

His eyes are drawn to the side of the screen. An animated ad plays in the lower corner, featuring the time-lapse satellite image of an enormous hurricane. The banner below it reads, "Has severe weather affected your life? Let's talk." He clicks on it innocuously.

A webpage loads in a new tab, entitled "The Brooklyn Project — Main Page." An impressively designed infographic loads with eye-grabbing animation. "If you knew the world had 50 years to live, what would you do?" reads a heading of red text. Underneath, an image of the globe appears, its saturated blues and greens fading into a putrid yellow and brown. It ignites dramatically, filling the page with digital flame. The video dissolves to black, and large words scroll over the screen. "There's still time to stop climate change. That's why we're here."

The display lifts, revealing a collage of faces, young people of all stripes. Despite the apocalyptic splash screen, the website now looks much like a standard university homepage — smiling faces, positive attitudes, and organized tabs of links. In the middle of the screen is an invitation: "Come see what we're up to. Next meeting is on Thursday at 8pm." An address below hyperlinks to an internet map; it's in Williamsburg, only twenty minutes away.

He grabs his phone excitedly. With furious fingers, he pounds out a text to Cooper and Shree:

Pando: u doing anything thurs night? 11:51am

Cooper replies first.

Cooper: Don't think so. 11:54am

Pando returns to the website, until, curious as to the definition of a term, he finds himself on an online encyclopedia. He travels from one link to another within the pages, taking him down a rabbit hole of endless discovery, each one related to the other in some degree, but the final course taking a meandering shape through the halls of human knowledge: North Atlantic oscillation, Cod fishing in Newfoundland, Factory ship, Fishing down the food web, Trophic level, Ten percent law, Digestion, Rectum, Anal sex, HIV, Bushmeat, Ivory Coast, Slave trade, Children in cocoa production, Cocoa bean, Hot chocolate, Churro, Lunch, Hot dog, Joey Chestnut. How he arrived here from his initial starting point is lost on him, but he's suddenly aware he's very hungry. He decides to grab a bite to eat.

A notification blinks on his phone. It buzzed without his awareness as he surfed the web in tunnel vision.

Shree: Nope. Date? 😬 1:38pm
Pando: something like that 2:06pm
Shree: Btw Mel says she thinks you should pay rent for as long as you're living on our couch 2:07pm
Pando: so i guess last night doesnt count then lol 2:07pm
Shree: Haha I guess not technically. Maybe you should join me more often 🙂 2:07pm
Pando: i mean if it will get me out of paying rent then hell yeah! lol 2:07pm
Shree: Let's have dinner together tonight. 2:08pm
Pando: sure, ill cook 2:08pm
Shree: Oh. Then nevermind. 😂 2:09pm
Pando: stfu ull like it. better than lunchables. 2:10pm

Navigating down the sidewalk is no easy feat at rush hour; the preponderance of bodies forms a slow-moving river of humans. Cooper is unable to break from its current, leaving him no choice but to travel at the glacial pace of foot traffic as he exits the subway station.

His body is hungry like it's never been before. All his income goes to rent, forcing him to scrounge muffins at work to stay alive. He's been rationing his food, spacing out his meals as far apart as his will allows before giving in to the gnawing pain. The cells of his biological machine are running on reserve power. His situation is growing dire, and he knows it.

As such, he's eliminated any and all superfluous expenditures of energy, such as exercise, partying, and traveling. This streamlining has left him a shadow of his former constitution. Efficiency has become the dominant preoccupation of his choices, with absolutely no energy to waste. His life has narrowed to three essential activities: school, work, and sleep. Anything beyond that is meaningless, simply because it is unavailable to him.

Inside the apartment, the refugee makes dinner for the roomies. Water is boiling in a large pot, blue flame curling up its silver sides. He hums to himself quietly as the bubbles roll.

Shree and Mel chill in the living room as they await their supper. Shree lies on the couch, her head propped up by the armrest, with Mel draped over her legs, their bodies intertwined. Shree is playing with blonde hair, twirling it sensually with her finger. She's staring at Mel's phone screen, held up for both to see.

"Ehh, not hot enough," Mel says, scrutinizing a picture of some dude in a black wifebeater with gelled-back hair. She swipes her thumb left across the screen, flicking the picture away and revealing a new one underneath. It features a man with red hair holding a French bulldog belly-up in his arms.

"Ew, ginger," says Shree.

"But he has a dog!" Mel exclaims. She swipes right. Another image appears, this time a woman with short curly hair.

"She's pretty cute," Shree admits.

"Yeah she is," agrees Mel, swiping right again. "When's that dinner gonna be ready?" she shouts to the kitchen.

Cooper swings the front door against the wall and kicks it closed behind him. He hauls himself up the ladder to his loft, shaking his head the entire way. He grabs his laptop off his mattress and returns to the living room, sitting down on the futon opposite Shree and Mel.

"Where you been, Coop?" asks Pando from the kitchen.

Cooper opens his laptop heatedly. "Work," he huffs. "I had a fucking day."

Pando turns off the stove and dumps the contents of the pot into a colander in the sink, and then tweezes the angel hair pasta onto four plates. He smothers them in meat sauce from a kettle on another burner, and sprinkles each pile with white crumbs of parmesan. He divides the garlic loaf into four quarters. Two at a time, he grabs the plates and heads to the couches.

"Here, Cooper, help me," he directs. "Dinner is ready, ladies."

Mel, who's still laying into Shree's body on the couch, pouts at the thought of having to sit up.

"Get off me, ho," says Shree.

"I don't wannaaaa," Mel replies indignantly. "Your body is so comfortable."

Shree knees her in the back sharply.

"Ow!" Mel shrieks as she sits up. "Jesus, relax."

"Here you go, Mel, fancy ramen," Pando says as he hands her a plate.

"Oh joy," she replies, taking the plate unenthusiastically. "You sure know how to impress a girl." She pulls out her phone and takes a picture of the plate of food. With the addition of a colored filter, she posts it online with the caption "roomeee dinner".

"This looks good, Pando!" says Shree. He finds the hint of surprise in her voice mildly insulting; it isn't hard to fuck up spaghetti.

Cooper returns with the other two plates. "Where did you get this anyway?" he asks. "I thought you were poorer than me."

"I am poor," Pando confirms. "And I dunno, I just found it in the cabinet."

Mel looks up at him. "Was it in the blue bag?" she asks.

"I think so," says Pando.

"Then it was mine," she growls. "Great. Glad you stole my food."

Pando stares at his plate dumbly for a second. "Oh…I guess I did, didn't I," he says half-heartedly. "My bad."

"Uhh yeah *your bad*. Thanks for asking me first, dumbass," Mel says.

"Calm down, I'll get you back," Pando promises. "Anyway, dig in everyone."

"Shouldn't we say a prayer first?" Shree asks. All eyes turn over to her skeptically.

"Umm I don't really *do* that whole prayer thing," says Mel.

"Yeah that isn't really my jam either," says Pando.

"Fine, I'll say it to myself then." Everyone stares at the freckled girl, Mel with a forkful of pasta held right up to her face, her mouth open with hair-trigger appetite.

"God, thank you for the food," Shree intones. "Amen."

They look at her contemptuously.

"There," she concedes. "Let's eat."

Mel stuffs the pasta into her mouth. They take a few bites in silence, nothing but the sound of chewing.

Pando swallows. "Dunno what you're thanking God for. Pretty sure this food was grown by humans."

"I'm pretty sure this food was *designed* by humans," Cooper adds.

Shree rolls her eyes. "Quit hating and eat."

Mel twirls a knot of spaghetti around her fork. "So speaking of getting me back, Pando, just curious: how long do you plan on staying here?"

Pando chews his food slowly. He glances over at Shree, who's blowing steam off a small pile of pasta sitting on the tines of her fork. She's avoiding eye contact.

"Uhh, well, I'm not sure at this point," Pando admits. "I don't really have a plan right now."

Mel smiles spuriously. "Well, as long as you're living with us and eating our food and using our electricity, I think it's only fair if you contribute toward rent while you're here," she opines.

"Well, I don't really have a job at the moment…" Pando again glances at Shree, who continues to avert her eyes.

"Like, am I right?" Mel interrupts, pointing with her fork. "How do you guys feel? Do you think he should have to pay if he lives with us?"

Cooper looks over to her, then to Shree, and back to Mel.

"Cooper? What do you think?" she insists. Pando turns to him.

The boy fumbles with his food. "Well, yeah, I guess that sounds fair. Although he doesn't like, have a real room or anything, so I don't think he should have to pay like an equal amount as the rest of us."

Shree nods in agreement and looks at Mel. "Alright, sure. Less than us, but not nothing," she agrees. Her eyes drift toward the ceiling as she does math. "How does ten dollars a day sound?" she asks, looking at Pando.

"Uh, yeah, sure. I can do that."

"Alright, that sounds reasonable," says Mel. "I'll collect the money every weekend. Sound good?"

They all nod.

"Sweet," she chirps. "So if Saturday was your first day here, then this coming Saturday you'll owe us 75 bucks."

"75 bucks?" Pando exasperates. "Where'd the five dollars come from?"

"From my pasta that's sitting on your plate," she admonishes.

He looks down at the spaghetti in his lap. "Alright, fair." He takes another bite. "How is it by the way?"

"So good," Cooper emotes.

"You're welcome," says Mel.

"By the way, what did you want to do on Thursday?" he asks Pando.

The refugee nods vigorously. "Yeah, I wanted to ask you guys about that. Mel, are you doing anything Thursday night?"

"Ummmm," Mel drones, uncertain. "I'll have to check. Why?"

He puts his fork down. "So I found this event online today that I think we should go to. Have you guys ever heard of the Brooklyn Project?"

They shake their heads.

"Apparently they're an organization dedicated to fighting climate change," he explains, fire in his voice. "They're having a meeting on

Thursday."

Mel scoffs. "Oh, please. Why would I want to go to that bullshit?"

Pando eyes her vexingly. "It isn't *bullshit,*" he snaps. "Really, we should go. I think it would be good for us. Especially you, Mel."

She squints judgingly at Pando. "I'm pretty sure I'd rather spend my Thursday doing literally anything else than listening to that hocus-pocus," she muses.

"I'd be down," says Cooper emphatically.

"Atta boy, that's what I like to hear," says Pando, punching Cooper lightly on the shoulder. "What do you think, Shree?"

She ponders it for a second. "Will it be as depressing as the stuff you said the other night?" she asks.

"Probably," says Pando matter-of-factly. "I mean I don't know. We'll only know if we go."

"Alright fine, I guess I'll go too," she capitulates.

His face alights with relief. "Mel?" he asks again insistently. "What do you say?"

"I'm taking the Nope Train to Fuck That Ville," she taunts.

Pando is more than a little hurt by her rejection. He doesn't even know how to react.

"Why can't you even listen?" he spews.

"Because I don't feel like it."

Pando shakes his head. "Whatever." He goes back to eating his food, lapping up red sauce onto his buttery garlic bread.

Mel scarfs down her slab of bread and finishes off the last bit of pasta on her plate. She rises from the sofa with a burp.

"Well that was good, Pando. Thanks for the ramen." She walks over to the kitchen with her dirty plate.

"Don't mention it," he replies.

She rinses off the plate and leaves it in a puddle in the sink. "Alright, goodnight guys," she says. As soon as her door closes, Pando turns to Shree.

"Can you believe her?" he asks in hushed tones. "Does she not even care?"

"Apparently not," says Cooper.

"Well fuck her then," says Pando. He continues to chew.

Shree, unhappy with this, pulls out her phone and texts Mel.

Shree: Come with me to the thing on Thursday 7:50pm

"At least the important people will be there," Pando enthuses, looking at his remaining roommates.

A loud hiss comes from the table. It's Shree's phone.

Mel: no fucking way 7:50pm
Shree: Pleeeeaseeee. It's not like you have anything better to do. 7:50pm
Mel: ...i hate you 7:50pm
Shree: Sooo yes? 7:51pm
Mel: omg fine. 😒 7:53pm

Shree smiles and locks her phone. "She'll be there," she announces.

Pando looks down with a smile at her slate on the table. "See, Cooper?" he gloats. "I'm not completely useless."

Neptune, the Mystic

The endless summer sales that stretched into late autumn have finally expired, the warm air of the city replaced by crisp coolness on this early December night. The four roommates pass down the streets of Brooklyn, Cooper out in front.

"Why am I here again?" Mel groans, shuffling along the filthy sidewalk, decorated with broken mailboxes and scraps of paper.

"Quit being a child," says Shree with indignation.

Cooper's phone navigates them to a bar with a large marquee sign that reads "Lucky Dragon No. 7." Peering inside, he can make out a smattering of tables, only one of them occupied. A gay couple sits at the bar, making light conversation with the bartender. There is nothing to indicate a meeting.

"You think this is the right place?" Cooper asks aloud, turning to Pando.

The Filipino looks through the window, witnessing the same scene as Cooper. He looks up at the sign. "Is it the right address?"

Cooper nods, showing him the screen.

"Guess it's just a load of crap after all," snarks Mel. "We should go home."

Shree walks stridently to the front door. "You guys are blind," she says, pointing to a paper taped onto the glass. It reads "Brooklyn Project meeting TONITE!!!" in green permanent marker. She opens the door for her roommates.

They scuffle in one by one. Pando walks up to the bar, leaning across the counter to get the bartender's attention. "Hey, we're here for the Brooklyn Project meeting?"

The woman looks up from her phone with a smile. "Oh, cool." She gestures to a black curtain in the rear of the bar room. "Just head on through there, go all the way to the end of the hall, and it's through the curtains on your right."

He slaps the bar counter loudly. "Awesome, thank you." He turns for the curtain, the roommates filing behind.

He parts the curtain determinedly. The corridor is long, illuminated by two naked bulbs at either end. The premises are auspicious enough for his taste; the worn hardwood floors and exposed brick give the hallway a rustic atmosphere in line with his expectations, and the curtain adds a twinge of mystery to it all. The sound of conversation grows louder as the four millennials approach the room. Above the door frame, a bulbous closed-circuit camera peers down at them.

Pando's intrigue is electrifying; finally, he's about to meet a group that actually understands what he's on about. They'll affirm that everything he's said is true, that he isn't actually crazy. He steps through the black velvet.

The first thing he sees is an enormous map of the globe posted over the far wall. It extends from the ceiling down to knee height, and is colored in varying splotches of deep oranges and reds. At its top, the Arctic Ocean is shaded a sickening imperial purple. The caption reads "Earth 2100" in thick block letters.

Before him are ten rows of benched woodgrain tables, arranged in pairs with a walkway down the middle. It leads to gray tables in the front, upon which are several ordered piles of fliers, brochures, and other informational materials. Brown cardboard boxes hold an assortment of pins, posters, and bumper stickers, from which several children are grabbing freely, hungrily collecting every variety: Keep It in the Ground; 2C or Bust; I'm an Alarm Clock; Right the Imbalance; Do We Really Want to Live *Here*?

The room is nearly full of people, with all tables occupied except the backmost bench on the left. Pando makes his way to the open seats, the other three trailing behind. They slide down as far as they can comfortably fit, but even then, Mel has barely enough room to support both of her cheeks. She grumbles unhappily.

This feels too much like church, she thinks to herself.

Cooper marvels at the turnout. The conversation is low but active, mostly confined to individual tables. The audience is comprised of people of all ages, from the few preteens sitting with their nuclear families to the occasional boomer. But the vast majority of the crowd tonight is millennial, a cohort of young people in every color and creed, the most diverse generation in history. A sensationally vibrant, endlessly fascinating pastiche of tastes, styles, and ideas. The millennial generation. His generation.

A mechanized whine fills the room, drawing Cooper's attention to the front. A large projector screen descends from the ceiling, slowly covering the ominous reds and violets of the large map on the wall. A searing blue light flickers on from the projector mounted above, fading to a white presentation slide:

BKP

Brooklyn Project

People continue to file into the room behind them, standing respectfully in the limited space in the back. Out of a corner door by the lectern in the front, a young woman enters the room. She dims the yellow lights from her phone, adjusting a virtual knob on the screen to half power. The children at the front table recognize that the meeting is about to begin, and cordially start back to their seats.

The dark-haired woman positions herself behind the podium with the authority of an army general. "Alright everyone," she begins. "Since we're more or less out of space at this point, we're gonna go ahead and get started a little early tonight."

Everyone has returned to the benches. With the exception of Mel, who is on her phone, every person in the room faces the front with anticipation, ready to consider the young woman's words. They've all come to hear her story.

"First of all, I'd like to welcome each of you to Brooklyn Project. My name is Nevaeh Brooks, one of the founders of BKP, and I'll be leading you through tonight's program."

Shree notices that Mel's still on her phone; she jabs her toe sharply into the girl's calf. Mel glares at her acrimoniously, but then softens, acknowledging her own disrespect. She powers down her screen and slips it reluctantly into her pocket.

"Just so you're all aware," Neveah continues, "some of the things you'll learn about tonight may be unsettling. We at Brooklyn Project want to stress that this room is a safe space, and if any of you feel the need to speak your mind about anything, then by all means, raise your hand and we can talk about it."

She looks down at the small tablet in her hand, checking off talking points with her thumb in the presentation app. The image on the projector screen automatically advances, with the new slide bearing the question, "What is BKP?"

"I'll officially kick off tonight's presentation by addressing the first question on everyone's mind — What is the Brooklyn Project? First let me explain a bit about who I am. As I said before, my name is Nevaeh Brooks, and I am a 35-year-old climate change communicator. How many of you have heard of climate change, or global warming?"

Every person in the room raises a hand into the air, even the small children.

"Phew, that's a solid start. When I started learning about climate change, I was so aghast that I couldn't sit still. I knew I had to dedicate my life to doing whatever it takes to stop the threat that it poses to everything and everyone I care about."

The screen advances, revealing an image featuring a quote over the face of Martin Luther King, Jr.

"As my idol Dr. King said, 'In a real sense, all life is interrelated. All people are caught in an inescapable network of mutuality, tied in a single garment of history.' If that's right, then this struggle is everyone's struggle."

The next frame appears, a real-time tally of the global human population, written as a ten-digit number that drives upward at an unflinching pace.

"And here are our ranks," she says, "the sum total of human beings on Earth."

The people in the room gawk at the figure, a staggering amount of homo sapiens just like them. Billions and billions of them.

She switches to a new slide, featuring a diagram of the human gut and an inset photograph of green microscopic bacteria. "But it goes much deeper than that, because we aren't just human beings. Right now, in your gut, there are more bacteria than the number of humans who have ever lived. In each one of you right now, there are ten times as many bacterial cells as there are human cells. In many ways, your body is only ten percent human."

The people in the room squirm uncomfortably in their seats. One girl rubs her belly vociferously, her mouth agape as she looks to her father in disbelief.

"Don't be grossed out by this, though," Nevaeh continues. "This is all perfectly natural. You've lived with these microbes in your body since the day you were born. You can't *live* without them."

The slide advances. A diagram of a food web appears, a series of arrows indicating the flow of energy from one trophic level to the next. The producers sit at the bottom: phytoplankton, plants, algae. They're eaten by the next level: fish, cows, poultry. Finally there's a human being, eating everything in sight, perched squarely at the top of the food chain.

"Of course, you depend on a lot more than just bacteria to survive. You need food: meat, fish, plants, and all the things that support it, like soil, air, and water. All the energy on Earth travels up the chain to you.

"So it should come as no surprise that when these things disappear, we disappear right along with them. Which is why climate change is such a big deal, because these things," she says, motioning to the plants and animals in the food web, "are starting to disappear."

The next slide appears; it shows three children laughing on a swing set, their mother beaming merrily behind them. The woman clasps her pregnant stomach, round with fetal life. She looks so bright, so divine, full of vitality and spirit.

"I'm the oldest of four sisters," Nevaeh reveals, "so I've always lived with a lot of responsibility, especially after my mother died when I was fourteen." She points to the angelic woman on the screen, an image now steeped in sorrow.

"She always used to tell me, 'Nevaeh, look after the children.'"

The next slide depicts four silhouettes leaping off a dock into a deep blue lake, emerald trees framing an orange sunset. The hands are locked in unity midair as their hair parts into wispy fans. Each figure gets progressively smaller in stature: a fully grown woman on the left, down to a small toddler on the right.

"So I did look after them," she affirms. "These are my sisters and me: I'm on the left, next to me is Jada, then Tierra, and finally little Natty.

"One day, not long after this picture was taken, my dad came home and sat me down on the couch. So I asked him, 'Daddy, what's wrong?' And he puts his arm around me and says, 'Your grandkids will never see fish in the ocean.'"

The people in the room stare at her unblinkingly, fifty pairs of captivated eyes.

"I looked at him in disbelief. I was like, 'Wait *what*?' And he explained that, based on how quickly they're disappearing, there won't be any fish left in the ocean in forty years."

She turns to the next slide. It shows a steep curve, the decline of fish in the ocean projected through midcentury, crashing to zero in a calamitous mass extinction.

"And of course I was stunned by this. How could it be true? Don't like, a lot of people eat fish? Aren't they gonna be hungry?"

She swipes to the following slide, a graph of bee and butterfly populations since the late 20th century. The trend lines decline precipitously, plummeting to its lowest on record in both cases for the most recent data points.

"And it isn't just the fish. This is a record of honey bee and monarch butterfly numbers in North America. As you can see, they're disappearing too. These species are key pollinators of our crops; together they account for over half of our food production. If they disappear, we don't eat, and if we don't eat, then we disappear too."

Tension grabs hold of the bodies in the room. Cooper bites his lip nervously, his sweaty brow betraying his crippling anxiety. This is not what he wanted to hear, yet the data before his eyes is difficult to ignore.

"The world wildlife population has halved in my father's lifetime alone. If the trend continues, the 21st century will be a century of mass extinction, and humans may be its ultimate victim.

"But these are only the symptoms," Nevaeh clarifies. "Does anyone in this room know the root of the cause?"

The screen clicks forward, revealing a graph of carbon dioxide concentration in the atmosphere over the decades. The level rises upward in an oscillating pattern, a perilous staircase climbing relentlessly at a faster and faster rate.

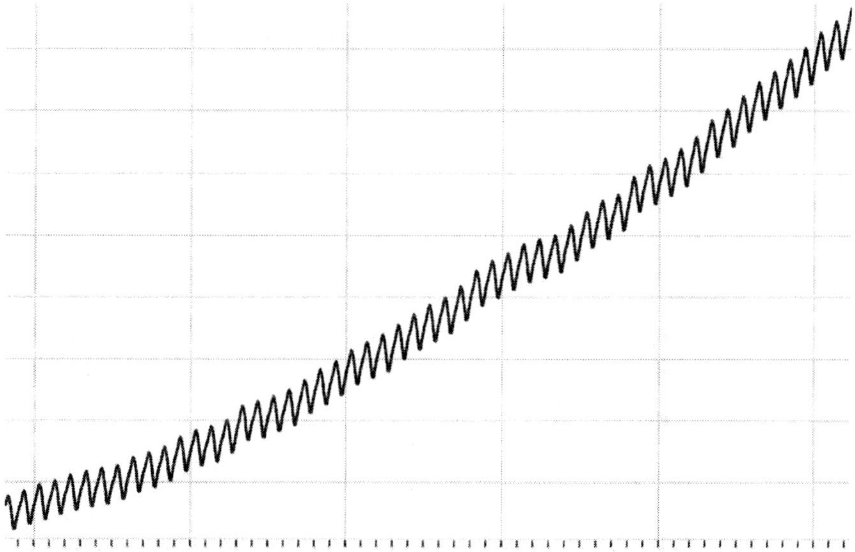

"Carbon pollution," she resounds. "As you can see, the amount of carbon dioxide in the air keeps rising; it hit a record this year, breaking last year's record, which broke the record the year before that, and the year before that. It's called the Keeling Curve."

She points to a rung on the graph. "Every day, this number goes up. The higher it goes, the hotter it gets. The Earth is already a full degree Celsius warmer than it once was, and before long it will be two. Right now we're on track for 4C."

The slide flips to another counter, this one in red font. It's the amount of carbon emitted into the atmosphere since the industrial revolution, its digits increasing so quickly they're a mere blur on the screen. In the lower corner is a timer, how much time is left before humanity burns a climate-killing amount of carbon. It's only a couple decades away.

"Here's where we are," she says grimly. "This is our carbon budget, and how much we have left to spend. Here's the counter. Here's the countdown." She readies her thumb over the tablet in her hand. Here comes the part no one wants to see.

Just then the side door opens; a head pokes its way through. "Sister," the blonde woman speaks, gently calling Nevaeh's attention to her. She nods downward curtly.

Nevaeh turns to face the crowd. "I apologize, but I have to step out briefly, if you'll all just excuse me one moment," she says. She joins the young woman and they disappear behind the door.

Cooper notes this curiously; it seems like an awkward interjection. But his mind wanders back to the counter on the screen, the numbers flying higher at an incomprehensible rate. Every cheeseburger, every escalator ride, every flip of a light switch sends more carbon into the air. It's inconceivable how many lights are on in this one city alone, let alone the whole world. No wonder we're all so screwed.

Nevaeh returns to the still-silent room, resuming her place at the front. "Sorry about that," she apologizes. "Anyway, what I was going to ask, and what many of you are probably wondering, is what happens if we don't act?" She brings her thumb to the touchscreen.

A simulated map of the globe is exposed on the screen. It animates itself forward in time, changing in shade from white to light yellow. A counter in the lower right corner measures the year; two more in the opposite corner measure CO_2 and sea level rise respectively. The top right corner has a counter for the human population and its malnutrition rate; the top left counter measures percentage of species extinct.

The years pass. 2020, 2030, 2040, 2050. The globe tints a deep orange over the continents, with the Arctic turning a dark crimson. 2060. 2070. The human population has peaked at ten billion and is falling calamitously. Malnutrition soars; the land bakes. Societies are suffocating in the extreme heat, now over 3C hotter than the 20th century average. The seas have risen fifteen feet, and ocean is encroaching on coastal cities. 2080, the land turns red; 2090, the Arctic is sullied purple. At last the map plays through to its final condition, the year 2100. It matches the map that greeted them on the wall, the sinister ghost of climate change yet to come.

The horror on the screen stares at Cooper. Manhattan is underwater. His home in Florida is too, along with the entire Miami area. The globe is stained an infernal red, like all the blood in the world had been spilled into great lakes across its surface. The worldwide wildlife population has decreased 85% since 1900; a third of all species have gone extinct. Fish populations have collapsed across the globe. Coral reefs are dissolving in the hot, acidic ocean. Global temperatures have spiked more than 4C above 20th century levels. The human population has crashed to two billion, most of them starving.

Shree looks at the screen in awe. *Hell on Earth,* she thinks to herself. *We're gonna pay for our sins.*

Nevaeh turns to face the petrified crowd. "This is business as usual. If we don't start treating our carbon addiction right now, this is where we'll be. Four degrees. And we're well on our way.

"What does it all mean?" she shouts. "What is this all about? Why should I care?" She pauses, looks around the room. "That's what some of you are thinking out there. Why should I care about this? Well let me frame the question like this: how many of you plan on living to midcentury?"

Most of the hands in the room go up.

"Then congratulations," she yells, "because you'll be alive to see this. Now how many of you knew this?" she raises her voice to the audience. "How many of you knew there won't be any fish left in your own lifetime?"

A smattering of hands goes up around the room. Pando's is one of them.

"Less than half," Nevaeh laments, scanning the responses. "That's unacceptable. It isn't your fault, though, because no one is telling you. But I'm telling you, because you need to know this. As a human being alive in the 21st century, you need to know this. So why should you care about climate change? I'll tell you why. Because it's about food..."

A quiet grumble tremors through Cooper's empty stomach. His

thoughts begin to drift, visions of Apocalypse forming from her words. *It's about food.* Bread lines. Famine. Starving children with bloated stomachs, fear and sadness in their eyes.

"It's about our farms turning to desert..."

The soil in Cooper's mind dries up like parchment, cracks running through the land like streaks of lighting as moisture is robbed from the ground.

"It's about our crops failing, our population crashing..."

More starving people. More riots. The end of organized society.

"It's about the complete collapse of the global ecosystem, and our economy with it..."

Armed revolution. Rogue states and extremism consume the world. A loose nuke is fired off. Oblivion.

"And it's about the fact that trying to feed ten billion people will be our own death sentence."

Nevaeh climbs powerfully onto the table in the front, exerting full jurisdiction over the minds and thoughts of her audience, stirring their hearts with a confusing mix of anger, sadness, and desperation.

"Ultimately, climate change is about us. It's about who we want to be, and how we want our kids to remember us," she thunders. "So why should you care about this? Because it is the worst thing that we could ever let happen, and it's happening right now."

Cooper's mind is sent reeling. His stomach is tight as a knot, a painful reaction to the map on the screen. It's the realization of his worst fear, what he's feared since the night of the tornado: nature is coming to finish what it started. And this time, it's coming for everything.

Despite the leaden atmosphere in the room, nobody stands to leave. Even the children are enraptured by the production before them. *This* is what they came to see: a climate warrior, with passion to ignite a revolution.

"So let's go back to your original question," she resumes after a pause for dramatic effect. "What is the Brooklyn Project? My sisters and I founded this organization with the support of our father. When he told me about climate change, my mother's charge came echoing back into my mind: 'Look after the children.'

"Well how can I look after the children when the world is on fire? When there's no food for their hungry mouths? When there's no clean water to drink, no electricity to power their lives? How am I supposed to look after the children *then*?"

The slide moves ahead. It shows four ebony hands piled on top of each other like a team huddle, a binding promise written in light.

"So my sisters and I made a pact — we agreed we would never have biological children of our own until the disaster of climate change has been irrefutably, resolutely averted. We would never be able to live with ourselves if we were to bring even one more life into a world like that. We won't let our kids grow up in a dust bowl amidst the worst economic calamity ever faced, civilization collapsing before their eyes like a nuclear holocaust. We just won't do it. We refuse.

"Instead, we've decided to adopt all the children of humanity, and we've taken it upon ourselves to fight for all of them as if they were our very own. Because if we don't, climate change will rob our kids from all of us."

The children in the room look at their parents. A father draws close his beloved daughter, a reminder that they're in this together, that he won't let her down.

"But we still have a chance to stop this," she assures them. "We've done it before. After Pearl Harbor, we as a nation worked together in the interest of self-preservation. We rationed food and energy. We retooled our whole economy. We did everything we could to ensure our survival, and we saved the world doing it. Because Americans can do anything when we put our minds to it.

"When we found out the Germans were working on an atomic bomb, we got to work immediately. We didn't waste time questioning the science. We got our act together, we started the Manhattan Project, and we invested the time and energy to create the necessary technology.

"It's the same thing here. Just as we knew then that Hitler was on the march, we know now that climate change is too. And it's gonna take all of us to solve it. It's gonna take a Manhattan Project. We need to do big things again. We need to save the world again.

"And that's why we created the Brooklyn Project. To avoid catastrophic climate change. To start a movement. It's more than just the name of our organization; it's an idea, and more importantly a plea. It's a plea to end the carbon pollution, immediately and for all time. It's a call for complete decarbonization of the economy, and a huge investment in clean energy infrastructure to provide for the human communities of the world. It's a call for America to ensure peace and freedom for the future, the way we once did. It's a call to greatness."

Another slide moves forward. It shows the old energies of the past — oil, gas, and coal — tossed aside in favor of the limitless potential of renewables like wind and solar.

"America can save the world again, if we choose not to destroy it first," she reiterates. "We need a full, concerted effort to deploy clean energy at every level of our economy. And the only way we're gonna succeed is if we start working on it *right now*, today.

"And that's why we want a Brooklyn Project. Like the Manhattan Project before it, a Brooklyn Project would be a matter of harnessing energy. But instead of using that energy for mass murder, we'd be using it to avoid it. The Brooklyn Project is about life, but as long we burn carbon, we're embracing death. It cannot continue any longer."

The presentation flips ahead again. It depicts young people like Cooper shaking hands with one another, working together to rebuild a nation.

"Aside from saving the world, a Brooklyn Project would create tens of thousands of jobs — jobs for people like us. It will usher in a golden age of electricity, a chance to do it right this time, a sustainable economy to last for all time. Decarb is not just essential for our climate, but our economy.

"This is an exciting idea; why aren't we following through with it?"

Nevaeh flips the slide. It depicts Joseph Coke on the screen, framed by names and faces of other government representatives that openly deny the threat of climate change. They're mostly old white men of the same political stripe.

"Because of people like this," she accuses. "This is Senator Joseph Coke, and he spends billions of dollars funding organizations that propagate climate change denial and oppose carbon regulation. These people refuse to acknowledge the devastating consequences of our carbon addiction, and they ignore their responsibility as our leaders to deal with it. They dismiss the fact that we know, with 95% certainty, that humans are changing the climate. If there's a 95% chance of rain, do you bring your umbrella, or do you pray you won't get soaked?"

At this point an agitated hand shoots up in the back of the room; Nevaeh pauses to address it. "Yes, miss, do you have a question?"

Mel stands to her feet. "Yeah, I have a question." She straightens her shirt with a downward tug, getting her big girl voice ready. "What about solving poverty? Or ending racial injustice? Or fixing education? Or making ourselves healthier?" she asks defiantly. "Why should we give our attention to glaciers when there are way more important battles to fight?"

Nevaeh smiles tenderly. "Those battles would be fought for nothing if we don't deal with climate change while we still have a chance." She folds her arms confrontationally. "You can't have utopia without Eden."

Mel frowns, slinking down in her seat. She doesn't agree, but that's nothing new.

"That's why deniers like Senator Coke must be opposed at all costs," Nevaeh affirms. "Let me ask you this: how many of you believe in gravity?"

All the hands go up, some more hesitatingly than others. It's hard to know what to believe anymore, now that all the empty promises of the world have been washed away, and the naked truth lies bare in front of them.

"Of course you all believe in gravity; it's all around you. It is what it is.

There's nothing incomprehensible about it. Well listen to me carefully." She huddles close to the audience. "As sure as gravity will pull you down, as sure as the Earth is round, as sure as the sun will rise tomorrow, *climate change is real.*" Her voice howls through the packed room. "It's real, and it's us. Don't ever forget that, for the rest of your life.

"You'll remember this day as the day you became aware of the greatest challenge mankind has ever faced. And just by being aware of it, you've joined the largest movement in human history: the movement to stop it.

"And that's why we need people like you to join us, to oppose ignorance and embrace understanding. There will always be deniers, people who refuse to see the truth; we'll never change their minds. But we don't have to. All we have to do is take back the conversation from them, and make them irrelevant. The rest of us aren't going to wait for the deniers to figure it out. The rest of us are gonna act. We will be alarm clocks, and we will shake the human race awake from this deadly sleepwalking. We will take control of our destiny, and we'll show the world that America can lead the way to a bright future once again.

"Those who deny climate change are more dangerous than Nazis. Once we were aware of the Holocaust and the mass extermination of human life, Americans didn't stand still. We didn't plug our ears and deny it. We recognized our responsibility to liberate the people in those camps. But the deniers would put humans in the incinerator again just to protect their unshakeable faith in the carbon economy. They're fascists, and they're shaping a fascist world. And we're not gonna let that happen."

She steps off the table with a deep breath. No noise comes from the crowd, not even a notification buzz. They're all staring at her, dumbfounded, steam wafting from their ears; she's blown everyone's minds. There are no words left to consider, because there are no words left to say. Until, when she speaks again, there are.

"That was a lot, wasn't it?" she asks genially, breaking the uncomfortable silence. "It isn't easy to accept that we've failed in our duties to tend the soil, to be stewards of our world." Nevaeh looks down with despair, as if she's bowing her head for a prayer. After a moment, she looks up. "How many of you believe in God?"

Hands. Lots of them.

"Climate change is an affront to God Himself, if you prefer to think of it in those terms," she says. "Allowing it to continue unimpeded would be mankind's greatest sin."

Shree's skin goes stone cold. *Mankind's greatest sin,* she thinks to herself. *Am I part of that?* The idea is too much for her to handle. The Apocalypse. Here it comes.

"So what's the good news?" Nevaeh asks finally. "The good news is this: we can solve this. We know what to do, and we have the ability to do it. And best of all, it's simple. We vote out the deniers once and for all. They do not deserve our vote, voting for them would be like voting for Hitler, with consequences far more devastating and far-reaching than any Holocaust. When the right people are making decisions, the rest will come. Simple, right?"

She scans the faces in the room. "I didn't say easy. I said simple. So simple that we can do it. If we can put a man on the moon, we can do this.

"So, alarm clocks, let's *do* this," she commands. "The evidence is all around us, as irrefutable as gravity and as abundant as air. Climate change is here. It's time to do something about it.

"Look around you," Nevaeh commands, pointing to the audience, "and take a hard look at all the people in this room now."

The heads in the crowd turn as they study each other with new eyes. Cooper looks around. Pando looks around. Shree looks around. Mel looks around. These aren't screens they're looking at; these are flesh-and-blood people. People like them, because all people are like them.

"Look at the diversity, the shapes, the sizes," notes Nevaeh.

The humans in the room study each other with pride and wonder. Every pair of eyes seems divine, a splendor of existence unique in the universe. And despite the differences, we're all bound together, and downright magnificent. We are human.

Nevaeh pauses on this note of singularity, this long moment where her captives drink up the dazzling creation that is human life. "We are a beautiful species," she reminds them. "And despite all that we've done, we are still a species worth saving.

"We're all roommates on this rock of ours, and we're gonna clean it up, because we look out for each other like good roommates do. We're a community, the human community. And as such, we'd like to borrow a page from Christianity by asking you all now to share in kiss of peace, and revel in the grace of humankind."

Shree looks to her roommate seated beside her. "Peace be with you, Mel," she says.

Mel, not knowing what the kiss of peace is, parrots the sentiment: "Peace be with you, Shree." She looks deep into her chocolate eyes, sweeter than bliss, and understands what Nevaeh is getting at. And she feels something stir in her that moment, something that makes her want to hug Shree like she's never wanted to hug anyone before, a deep admiration and outpouring of affection.

The others in the room repeat the statement to each other, pairing off one by one to wish for peace, in every way, shape and form: handshake, hug, kiss, fist bump. The twins engage in their secret dap. They all find a connection, and they feel the ties that bind.

"Peace be with you, buddy," says Pando, extending a hand to Cooper.

"Back at you, man," Cooper reciprocates. "Peace be with you."

Pando kisses Shree on the cheek, and whispers into her ear. Then he reaches his arm over to Mel. "Peace be with you," he offers insistently.

The blonde smirks and grabs his hand in an exaggerated shake. "Peace be with you too," she says. After the noise has quieted, the aggregate of humanity settles back into their seats, refreshed and reminded of just how great people can be.

"So here we are," says Nevaeh. "The clean revolution, right here in this room." She mounts the table again and stretches out her arms in a show of solidarity. "We're calling all of you to join us. We're calling for the modern day Paul Reveres and George Washingtons, the bold leaders who will spread the word and lead us to victory. We're calling for the Benjamin Franklins and Thomas Jeffersons to think our way out of this problem, to create a new kind of society that plans on sticking around for our children. We're calling for the Rosa Parks of the decarb movement, someone with the bravery to stand up to injustice. We're calling for the Abraham Lincolns to overcome political stalemates and simply make things happen. We're calling for the FDRs and Churchills to save the world from evil. We're calling for the MLKs for their oratory and vision. We're calling for the JFKs for the moonshot ambition and determination. We're calling the best people in America to get this done. Because really, we're calling for a miracle. And until we get one, it's up to us. We're calling for understanding. We're calling for action. We're calling for a Brooklyn Project.

"Thank you."

She steps down.

The applause is immediate and deafening. It was a life-changing experience; a moment in history. After basking in the glow for a moment, she quiets the crowd and begins the Q&A. They clarify the talking points and discuss their feelings. The situation isn't good, but it could definitely be worse. As they wrap the meeting, the people in the room avow to do the right thing.

"For those of you who wish to get further involved, you should sign up for our email list so you can keep abreast of whatever important events we have coming up." Nevaeh brings her hand down on the spreadsheet on the front table. "We're looking forward to the divestment protest at Columbia University in May, and we intend to show up with a small army."

Cooper's brain flashes recognition, the same divestment protest Natasha's environmental club is working on. This really is an integrated movement.

"I think we had a good session tonight, so I'm gonna call it a wrap, and wish each of you the best of luck," Nevaeh says pridefully.

The people rise. They're shaken, a little humbler, but ultimately rejuvenated, given a fresh look at the world. Above all, they feel more than a little hopeful. There will always be enough people doing the right thing, and the meeting tonight confirms it. Humanity is too big to fail.

Pando gets to his feet almost immediately, bidding to circumnavigate the crowd. He exerts his way to the front, eager to speak with the legendary speaker.

The bench shakes under Cooper; his body is shivering slightly, and he isn't even aware of it. Or maybe it's Shree that's shivering? They both are, trembling inside and out. Despite the positive ending, they both feel like they've witnessed Armageddon.

Pando waits in line to speak with Nevaeh. A single father with a young girl stands in front of him, eager to tell her how important her words were to them.

"I can't thank you enough for this," the man tells her, holding his daughter's hand. "It's really given me hope."

She smiles politely, and stoops down to the little girl's height. "I'm doing it for her," she reminds him, rubbing the girl's brunette head tenderly. She smiles as the two walk off.

Pando advances to the front. "That was awesome," he says. "You nailed it."

Nevaeh grins enthusiastically, happy to see another young adult inspired by the presentation. "Thank you! I worked hard on it."

"Everyone should listen to you." Pando glances at 22nd century Earth, lamenting the disaster it portends. "If we could just forcibly stick a light in people's brains, then they'd see," he says, pointing to the map.

"Well that's why you're gonna help me," she replies.

Pando nods. "I brought my friend so he could understand," he explains. "He's just a boy, doesn't know much about the world. But I think you really sold it to him." He emotes boisterously, his zeal coming through with every word. "Is there any way I could maybe get involved in what you do here?"

Nevaeh smiles, gesturing toward the table behind her. "Well here, you can join our mailing list."

Pando's heart sinks; he was hoping for something more tangible than a weekly email blast.

"Okay, sure," he says timidly. "Are you guys hiring at all? I'd love to do as much as I can."

She studies him for a moment, recognizing that he may have the ingredients they're looking for.

"You know, there just might be something you could do for us," she begins. "I'll have to talk to some people first, but there's a project we're working on that we could use some extra hands with. Maybe that's something you'd be interested in."

Pando's face lights up with titillation. "Yes! Absolutely!"

"Great," she replies. "Give me your email." She hands him the clipboard with the spreadsheet. He grabs the pen and scribbles in his contact info.

"Cool," she says, glancing down at his chicken scratch. "Alright Pando, we'll be in touch."

The refugee departs with a smile. As he turns to leave, Nevaeh marks a crude star next to his name.

He rejoins his roommates by the hall. "Let's bounce."

The four millennials travel home in complete silence. The train rattles loudly, a missile whizzing beneath the metropolis above. They sit next to each other, staring out ahead, none of them ready to acknowledge what they've just been through.

Mel is uncomfortable by all this quiet. She notices a semi-attractive man sitting on the opposite corner of their seat bank. He's only a seven, but even a seven would be worth talking about amidst the palpable tension between the four of them in this moment. Never has she seen her roommates so wound up. She pulls out her phone.

Shree's pocket buzzes. She's barely lucid enough to notice; she's still thinking about her role in the impending doom. She checks the notification anyway:

Mel: check out the hottie in the fedora. what do u say? 7? 10:16pm

Shree locks her phone dismissively and puts it back in her pocket. No one cares how hot that guy is right now. The world is fucking ending.

Mel, noting Shree's scorn, gives up on conversation. She unwinds her white buds and plugs them into her ears, looking to forget the inconvenient truths of her life.

Cooper is staring into his lap. Even though the battle is far from lost, he still feels an overwhelming sense of futility — maybe it is too late. The task is enormous. BKP seems primarily interested in political action, and if Pando's right, then the world is lost, because politics has failed. The stalemate continues, and time runs out.

Look at them all, Cooper thinks to himself, marveling at the extreme dissociation of everyone riding in the car. Each person has headphones in their ears, eyes on their screens. All around him sit closed-minded humans, ignoring anything and everything that doesn't tickle their immediate interest. In a world where information flows all around, when the air is alive with swarming data streams that can be accessed from anywhere at any time, when a human being could know anything that has ever been known from the slates in their hands, they just wear their headphones and crush candy, blocking it all out. They only listen to what they want to hear, so that reality for them is exactly what they want it to be. Humans have finally found a way to control the world — not by learning all there is to know, but by actively avoiding it.

They depart the train single file, Mel pulling ahead with quicker pace. She readies her keys for the moment she reaches the door. They walk into the building like weary ducks in a line.

"Thank God we're home," Mel sighs, pulling out her earplugs and spooling them around her phone.

"Yes, thank *God,*" Pando mimics. "So what did you guys think?"

"Waste of my time," she replies as she starts for the ladder.

Pando pays no mind to her. He looks to Cooper for an honest answer instead.

"It was thoroughly terrifying," he says.

Shree shakes her head, unable to articulate her feelings. "I just want to go to bed," she says.

"Agreed," says Pando. They all make for the ladder, climbing up in quiet. Mel closes her door without so much as a goodnight. Shree grabs Pando's hand tenderly and leads him into her room.

"Night, Coop," he says to his young friend as Shree pulls him behind the door.

Cooper nods to himself, sullen. Entering his room, he pulls off his shirt and collapses onto the mattress on the floor, drained of all enthusiasm for the future. He turns off the light and stretches out on the bed in darkness.

I'm gonna die, he thinks to himself, truly understanding it for the first time in fifteen years. A clock ticks angrily somewhere in his mind. Suddenly the world looks totally different, and he begins to realize, slowly, that despite all our good intentions, we were wrong. The mirage of the carbon economy hasn't amounted to anything more than a death sentence. Like a rogue cancer, humans are killing the body of life.

And just in accepting it, Cooper begins to understand that it all actually makes sense. This isn't some cruel joke. Climate change is real. And it's going to suck.

ANTHROPOCENE

Time to Pretend

The founding team of the Brooklyn Project sits relaxingly around their briefing room table. Jamal reclines in his chair, doing math on his clipboard. Nevaeh sips a cup of green tea that steams on the table. Tierra types on her phone, Jada by her side. They all have tablets in their laps.

"So what've we got tonight," Jada asks the group.

Nevaeh flips through her screen. "I found a few good matches," she says. "But I want to start with one in particular."

She taps a button to share her screen with the other tablets. She pulls up a still frame of video taken from the camera above the entrance to meeting room, situated as to record every face that passed through the hallway last Thursday. It features a tall Asian man with knotted hair.

"And who is he?" asks Tierra.

Nevaeh flicks a document onto the screen, complete with government identification pictures, biographic statistics, and a comprehensive personality profile.

Tierra reads it aloud. "'Bayao Pandong Lamar. 24 years old. American-Filipino.'"

Jada looks closely at the document. "Says here he was recently in Kabubwason."

The sisters look up to their elder with intrigue. Jamal lowers his clipboard.

"A refugee?" asks Tierra.

"Likely," Nevaeh reports. "Scores high on radicalization, as well as virtue, loyalty, and intellect."

"'Highly eco-conscious,'" reads Jada.

Jamal looks at the file from his tablet. "I think he's good. We should pursue him."

Nevaeh signs her name over the document and flicks it aside. She returns to the camera still.

"What about this one?" asks Tierra, considering the image of the four roommates, their names labeled. "This 'Sharifa Morgan Dunlap.' She good?"

Nevaeh shakes her head. "Nah, too unstable. But the white kid, on the other hand, shows a promising balance." She points to Cooper on the screen. "I crunched his numbers. He's similar enough to what we're looking for, within most of our target ranges."

"I know him," comes a voice from one of the Sisters. The others turn to her.

"You do?" asks Nevaeh. The girl nods, eyes closed.

"Do you think he'd be useful?" Jada presses.

The woman shrugs. "I can't be certain, but my guess is yes."

Shree is teaching a science lesson her students don't understand.

"And so the continents are drifting apart slowly," she concludes. Glazed eyes stare back at her, completely lost. "Any questions?"

The class sits in silence.

"Any at all? Now's the time to ask, boys and girls." Shree is keenly aware of the miserable state of her students' grades; they're just not grasping the material. No matter what she does, they don't understand.

"Please ask," she says desperately. "Please. You need to understand this."

Empty looks are all she gets.

She sighs in frustration. "If you guys can't understand basic geology, how are you ever going to understand climate change?"

A hand shoots up from the back of the room. It's little Alyssa, demonstrating once again that she's one of the brightest students in class. At least Shree knows she always has an audience in her. She calls on her with a point.

"What's climate change?" the girl squeaks.

Shree peers into the child's big, round eyes. That's a question right up there with "Where do babies come from?" on the list of hardest things to explain to your kids. She looks around uncomfortably. The other eyes in the room have suddenly sprung back to life, hungry with curiosity.

"It's..." the teacher begins, before breaking off. She looks across the room at the girl again. They don't break eye contact for a small eternity.

She's just another victim, she thinks, pondering the child. *They all are.*

"Don't worry about it," she says at last.

The bell rings unexpectedly, prompting the students to their feet.

"Now wait," Shree says, scrambling through her lesson book. "We didn't cover a lot of things, like magma and lava. Basically, magma is underground lava. Okay? If you see that on a test, that's the answer." She closes the book amidst a sea of blank stares. "Alright. You can go."

The students grab their bags energetically and walk out the door in bunches.

Shree is surprised it's already noon; time marches faster and faster these days. The clock ticks. Another 350 tons of carbon in air. Only 147,000 tons to go until another extinction. Tick, tick. Death, death. On and on. Forever and ever.

She heads to the teachers' lounge and grabs her Lunchables out of the fridge. She wants to eat outside today; she needs the fresh air.

She exits the building into the cityscape beyond. The buildings around her twist into giant blades that rake the sky. They look violent, menacing, taking on a whole new uncertainty in her eyes. Once upon a time, buildings were just places people go to work for the day. But what are they really? The teeth in the saw that fells the tree of life? Instruments of mass destruction? She isn't sure anymore.

She walks along the sidewalk to a park by the school. The once safe and familiar landmarks now look completely alien and dangerous, as hostile as a tiger on the savannah. With the instinct of fear in her head, her eyes scan the skyline with distrust. Anything spewing carbon is a deadly enemy, and these great pillars of glass are no exception, peopled by mindless sheep oblivious to the invisible danger rising all around them.

. . .

Cooper draws breath sharply, waking in a cold sweat. The boy looks around himself uncomfortably in his lightless room. Thirst beckons; his throat is drier than California. He exits his tiny loft to pour himself glass of water.

Down below, he finds Pando awake on the couch, his ghostly face illuminated by the glow of a computer screen. A small plate sits next to him on the table, its brown crumbs barely visible on its surface.

"You're up late."

Pando glances in his direction. "It's only 3:30," he says.

"What are you doing?" Cooper asks as he plods into the kitchen.

"Just writing," he says. "Also I want to be up in case I hear back from BKP."

Cooper laughs in amusement. "I don't think they're gonna be sending any emails at 3:30 on Sunday morning." He returns to the living room with water in hand.

"I can't sleep," the kid says, clutching his plastic cup.

"I can see that," Pando replies tersely.

"I keep having these weird dreams."

Pando expresses no interest, but Cooper keeps talking anyway.

"Each time I'm on some huge sinking ship, I think it's the *Titanic,*" he explains. "And I'm running for my life, cuz, like, you know."

Pando lifts an eyebrow, his scar gleaming in the darkness along with his scintillating gray eyes. "That sounds pretty scary," he says without taking his eyes off the computer. He moves his finger rapidly across the touchpad, highlighting a string of text and deleting it. From Cooper's vantage, the page looks like a poem of some sort, even rows of words with breaks in between.

"Yeah, it is," Cooper agrees. "Especially since I can't swim. It's like my greatest fear."

Pando gives him an incredulous look. "You don't know how to swim?" he asks.

Cooper shrugs. "Nope. And I grew up in a *lakehouse* until I was six, so it's not like I didn't have the chance."

"Now's probably a good time to learn," Pando says, chuckling to himself as he contemplates the rising sea.

"I dunno," Cooper says, rubbing his chin. "I haven't been swimming since then, since that day."

"And what day is that."

"The day nature ruined my life."

Pando's head cocks up. "I don't think that day has come yet," he says.

Cooper is unamused. Now isn't the time for end of the world jokes.

"So what happened?" Pando asks finally.

The boy folds his hands together. "Well," he begins, before halting. He hasn't told anyone in a long time; this might be hard. "Actually, could you maybe close your laptop?" he asks humbly.

Pando considers the screen, then pulls it down with his finger. The light is extinguished; darkness blankets the room. Cooper takes a deep breath.

"So when I was six," he describes, "my house was hit by a tornado. My dad and sister and I survived, but my mom didn't." He pauses for a moment, trying not to let emotion grab him.

"She used to swim with me every day," he reminisces. "We used to have so much fun. But then we moved from Kissimmee closer to Miami, and I stopped going in the water."

He thinks back to that horrible day. The noise above as he sat in darkness in the basement. All alone with the radio.

"They said it was an El Niño," he recalls, "that maybe an El Niño caused the tornadoes."

More silence in the dark. Is Pando even listening? Did he fall asleep?

"Anyway," Cooper closes, "that's what happened."

Pando shuffles audibly on the couch. "I'm sorry, man," is his overdue reply. He was listening intently, as always.

"The weird thing is, I had a dream about the *Titanic* that night too," Cooper says.

"That's crazy," says Pando.

"I know."

Another space of silence. Silence and darkness. Then Pando speaks up.

"You want to hear something?" he asks.

"Sure."

"Can I open my computer?"

Cooper nods; Pando understands.

The clamshell emerges in a beam of white light, eight million phosphorescent pixels strong. Pressing a button on his keyboard, Pando plays an audio track that he's composed, a layered hip hop rhythm with tight synthesizers. After finding the beat with his head, he begins to rap:

> Here stand the youth, we ready to fight
> The carbon in our lungs, the end clearly in sight
> But the old men say no, they'd rather just talk
> While the rest of us sit still, helpless, and balk
>
> And all their education
> And their college degrees

And all their negotiation
Can't keep us at two degrees

Cooper's eyes are large with stupefaction. "Damn, Pando," he mutters. "That was dope."

The refugee smiles weakly. "Thanks, bro. I've been working on it for a few weeks now."

"That was like, really powerful," Cooper praises. "Maybe even revolutionary."

Pando laughs. "Yeah, well, it's pretty much all I can do at this point."

With that comment, Cooper's thoughts return to the fatalism of his life, and the enormity of dealing with the world's terminal cancer.

"What about BKP? They're doing something," he points out.

"Not really," Pando opines. "I gotta admit, I was hoping there would be more to it. But their solutions are political, and you know that's not enough. Not at this rate."

Cooper frowns. "Yeah, I guess you're right," he laments.

Pando rises to his feet and grabs the dirty plate off the table, taking it over to the kitchen sink. "Of course I'm right," he replies. "I'm honest. Honesty is about telling the truth, and truth is all I know." He smiles cockily into the enveloping blackness; Cooper smiles back.

"I wish you weren't so damn right all the time," he says.

"It's a blessing and a curse," he explains. "The blessing being at least I'm aware of the reality of our lives, the curse being that I have to find a way to live with it." He puts the plate in the sink, gives it a rinse, and leaves it there in the puddle.

"Anyway, I'm heading to bed, man." He walks over to his roommate and rubs his hair affectionately. "I'll see you tomorrow."

Cooper raises his glass as Pando makes the climb to Shree's room. He finishes the rest of his water and heads to the sink. Seeing the plate submerged as it is, he puts the empty cup on top of it, yawns sleepily, and shuffles back to bed.

The following morning, Mel is finishing a bowl of cereal in the living room. She goes over to the kitchen to clean her dish, only to find the dirty dishes her roommates left overnight. She frowns, adding her bowl and spoon to the pile.

She's still waiting for Shree to return from church. Bored, she sits back on the futon and starts playing on her phone, distracting herself with a barrage of texting, browsing her friend feed, watching stupid videos, whatever. It doesn't matter. She swipes through it all, everything and anything that scrolls into view. The internet is like a trail of candy for her, leading her from one rewarding nugget to the next, strung along by who knows what, extracting no larger meaning, plotting no coherent course. It's all just a blur of words and images.

She picks out a post from Shree in her feed, posted less than half an hour ago:

> Shree Dunlap: I brought you into a fertile land to eat its fruit and rich produce. But you came and defiled my land and made my inheritance detestable. Jeremiah 2:7 11:47am

Another Bible quote; she must be coming home soon. Sure enough, within minutes, Shree arrives. Mel looks up from her phone as soon as the door opens.

"Hello darling," she says, getting up to greet her with a hug. Shree seems tired and sad, limp in her arms. Mel looks at her with concern.

"Are you okay?" she asks. "It looks like you've lost your pluck."

"I'm just tired," Shree lies. She slouches past and kicks off her shoes onto the floor.

"What are you doing today?" asks Mel.

"I'm not doing anything today," comes Shree's reply. Her one shoe isn't coming off easily, forcing her to pull it off with her hand.

"You want to do boozy brunch?" Mel asks eagerly.

"Not really."

"Oh come onnn," she whines. "We haven't gone in weeks."

"No," Shree repeats irritatingly, mounting the ladder. "When I say I'm not doing anything, I mean I'm *not doing anything.*"

"Whoa. Alright then." Mel watches her roommate climb up the itchy rope; she isn't normally this crabby. "Are you on your period?"

Shree stops her ascent to glower at Mel below. "You know we're in sync," she says, resuming the climb.

Mel's eyes follow her as she traverses the narrow walkway. "Oh, by the way, tell your boyfriend he owes me money," she calls. She watches the door close.

In her room, Shree finds Pando still sleeping in her bed. She rustles him awake with a foot. He squirms uncomfortably in the warm bed.

"Fuckkk youuu," he mumbles. "I was having a good dream."

"Oh yeah?" asks Shree, pulling off her socks. "So I was in it?"

He smiles and buries his head into the pillow. "Something like that." He yawns and stretches in bed, then sits up, assertively defying the allure of sleep.

"Have you heard from BKP yet?" she asks.

"Nah," he replies glumly. "I've pretty much given up on them. It's been a few days." The disappointment in his voice is apparent.

"Don't lose hope," Shree says.

Unswayed, Pando shakes his head. "It's not like they're really doing anything anyway. Just another group of hippies trying to talk their way through climate change. Talk isn't enough."

Shree rolls her eyes. "You're a real delight to have around." He flips her the bird and smiles. Amused, she collapses down onto the bed next to him, running her fingers through his hair. "Did you pay Mel yet?" she asks.

His eyes go wide. "Oh shit," he says. "Imma get to it later."

She sits up in surprise. "Are you for serious? She's gonna be tight."

"She'll get over it," he says.

Shree continues to play with his knotted hair like tassels to a cat. "When are you gonna get a job?"

His eyes roll back in his head, and he collapses onto the bed dramatically. "Ehh, I dunno. When I find a reason to." He frowns into his pillow. "I was gonna apply to jobs this morning, but I decided to sleep in instead."

She pulls the pillow out from under him and beats him over the head with it.

"You so lazy," she says indignantly. "You're the king of procrastination."

Pando laughs. "Well hey, I'm a millennial. We're all supposed to be lazy. Get used to it. We're the future."

Shree snickers at the thought. "So why do today what you can put off till tomorrow, is that what you mean?"

He covers his face with his hands. "No, more like why do today because there is no tomorrow?"

She hits him over his head again. "Cooper's right," she says critically. "You aren't even trying."

Pando sits up in bed. "Because I'm tired!" he bellows.

"Tired?" Shree asks. "*You're* tired?"

"That's right," Pando says.

"What are you tired of?"

Pando exhales loudly. "I'm tired of dealing with people and their infinite stupidity," he explains, channeling the experience of an old man. "I'm tired of giving a damn about this world when no one else does."

Hearing this, Shree realizes something fundamental about the man: maybe when his life feels like something that might actually last, he'll start taking care of himself. But until he sees credible evidence that his life is worth something, he has no real motivation to do anything.

"If you don't at least get a job interview by Tuesday, I'm kicking you out," she says firmly.

He looks up at her mistrustfully. "You don't mean that," he says.

"I'm dead ass."

His eyes narrow, but her rigid stare isn't yielding. Backing down, he breaks eye contact to rise from the bed. He collects a pair of socks, a shirt, and some pants. He pulls the socks onto his feet and starts getting dressed.

"What are you doing?" she asks.

He doesn't speak. He yanks his pants up, affixes them to his waist with a belt, and slides his arms into the shirt sleeves. He slips his cellphone and wallet into his pocket.

"Where are you going?" Shree asks, anxiety in her voice.

His filthy shoes sit tied by the door. He pushes his feet into them with minimal effort, opens the door, and steps out. He turns his head to look back at her.

"To find a job," he says. He closes the door behind him.

Organized tributaries of people stream through the underground ramps and passageways of the Times Square subway station. Tourism is at its holiday season peak; the foot traffic is at least six million strong. So many people. So many mouths to feed.

Pando is among them, transferring to another train. He pulls ahead of the throngs, deeming them too slow for his pace. He opens his stride as he travels down a ramp, moving past a section of walkway staked out by a man proselytizing in casual dress. In his hands is a stack of paper brochures, some of which have variously-colored paperclips affixed to their tops. Large white picket signs are spaced evenly down the path, each one written with various quotes. If Pando hadn't been a climate warrior, he may have dismissed this as nothing more than a religious display aimed at converting nonbelievers. But noting the large poster of 22nd century Earth on the wall, he recognizes it as a BKP informational setup, one of their many outreach programs.

"Climate change is here," the man promises. "Nobody is paying attention."

The people walk on by, most of them on their phones. Nobody is paying attention.

Pando, drawing nearer to the man, listens but doesn't look. He knows climate change is here. He's still waiting for more than talk.

The BKP spokesman wears a white headset that feeds him complicated streams of information as it analyzes every passing face. Sifting through the infinite data of digital life, he is able to identify anything and anyone in his field of view.

Pando steps nearer. The headset identifies him immediately, flagging him yellow in the wearer's view.

"Pando," the man says, thumbing over his stack of pamphlets to the yellow paperclip.

The refugee halts abruptly at the sound of his name. A fellow New Yorker, walking not far behind, slams right into his back.

"Watch it," the man says, bouncing off Pando's shoulder. He doesn't notice; his focus is on the BKP rep before him, who's holding out a pamphlet.

"For you," says the man.

Pando steps forward. "Me?"

The man nods stoically, the pamphlet extended inexorably.

He grabs it from him; the man smiles. With no more interaction necessary, he resumes his sermon.

"Climate change can still be stopped!" he shouts as the people walk on by. Nobody is paying attention.

Pando continues to the train, pamphlet in hand. As soon as he takes a seat, his eager eyes begin to read. It's a handout of the same design as those from the info session. It has graphs, web resources, and lofty rhetoric. But tucked underneath the yellow paperclip, a tiny square of paper has been folded inside. He opens it hurriedly.

He notes the signature in the bottom as he pulls out his phone.

Pando: yo where u at 12:39pm
Cooper: Work. Why? 12:41pm
Pando: bkp tomorrow. 5. 12:41pm
Cooper: I've got class ☹ 12:44pm

This is a frustrating response. Really? Really Cooper? You have class? Please.

Pando: when u coming home tonight 12:45pm
Cooper: Like 8. 12:45pm

Satisfied, Pando locks his phone and slouches lazily into his subway seat.

Cooper's earbuds fill his world with music as he walks through the door to the apartment. There's no sign of Pando on the couch, so he removes his earphones and begins up the ladder. As he collapses onto his bed, he hears hip hop music emanating from Shree's bedroom.

Cooper: Just got home. 8:13pm

The music through the wall ceases, and Cooper can hear the bed springs groan. Two feet appear through the slit of light at the floor. Three raps.

"Come in."

The door creaks open, revealing a dark figure imposed against the white light. There's a slip of paper in his hands, and urgency in his posture: he's straighter than usual, not as slouchy. He holds himself up with purpose.

"I don't say this often," Pando whispers into the night black room, "but I was wrong."

He shuts the door behind him.

"Maybe there is something we can do."

Shree and Mel parade through the front door, all giggles. Both of them are more than a smidge intoxicated. Mel flings the door shut, stumbles inside, and nearly trips on a pair of shoes that have been left carelessly by the bathroom. With angry eyes, she kicks them to the side.

"Pando!" she screams at the top of her lungs.

Shree reacts with surprise. "Jesus, you don't have to yell."

"Yes I do," she says sourly. "That fucker owes me money."

Pando opens the door from Cooper's room. "What do you want?" he calls innocently.

"Get down here," she snorts.

Reluctantly, Pando heads downstairs. "What's up," he says, facing her.

"Are these your shoes?" she asks.

"Yeah, why."

She bunts them over to him; one shoe skips wildly over to the left. "Take care of them," she orders. "Also, I don't know if you've noticed, but it's Sunday, and you were supposed to pay me yesterday. Do you want me to start charging you late fees?"

Pando grimaces and hangs his head. "I know, I'm sorry. I have the money." He thrusts his hand into his pocket and pulls out 85 bucks. "Here. Sorry it's late."

She looks at it for a second, squinting into his eyes of pewter as she snatches the money out of his hand. "You need to be more responsible," she snaps. "I shouldn't have to hound you for rent."

His face heats, but he calms his nerves with measured control. "It won't happen again," he avows.

"It better not," she says, pushing him aside drunkenly to climb the ladder. He looks over to Shree, shaking his head disapprovingly.

I told you, she mouths. They follow her up the ladder; he rubs her back fondly as she walks into her room.

He returns to Cooper's room and closes the door. He resumes packing the bowl of his bubbler, then snaps his fingers demandingly for Cooper to hand him the lighter.

"I don't think I'm gonna go," says his roommate.

Pando exhales a plume of smoke with a cough. "Dude, why the hell not?" he gasps. "Clearly they want you there." He hands Cooper the pipe and lighter.

"Yeah but why?" the boy asks, bringing flame to the bowl. "What could they possibly want with me? You're the one who cares so much about this stuff." He takes a hit.

"Does it matter why?"

Cooper blows a large white cloud into the air above him.

"Listen, you need to come," says Pando. "They might turn me away if I go without you." He grabs the pipe from Cooper.

"Dude I've got class," he reminds him.

"As if your degree matters!" Pando shouts. "This is your chance to do something, to help the world right now. You're really gonna walk away from this so you can finish the degree you don't even want?"

The boy watches him medicate with fragrant bud. They're both fried.

"It's one class," Pando bickers again.

Cooper draws a breath of hazy air. "Okay, fine," he relents. "I'll go with you."

Across the Universe

The two young men have returned to the Williamsburg bar. The bartender on this afternoon happens to be one Tierra Brooks; Pando leans over to her.

"Hey, is Nevaeh around?" he asks.

Tierra sizes them up with a quick glance. "I believe so. You need to see her?"

Pando thrusts his hand into his pocket. "Well, I'm not really sure," he explains. "I got this." He hands her the cryptic note.

Tierra looks at it with intrigue, then turns back to the two humans standing before her. "I'll be right back," she says. She disappears behind the back curtain, leaving the two roommates to chill for a minute.

Cooper stares at the tree rings in the wooden boards of the bar. Each tiny ring in the grain is an entire year. He traces his finger over one of them tenderly. What year does this one represent? Was he just a toddler, running around by his mother's side? And this ring — only a few inches away, yet years in temporal distance — why is it so thin? Was there a drought that year? How old would he have been then?

Tierra pokes her head through the curtains. "Follow me, guys," she says.

The three of them walk through the velvet and down the hall, entering the same large presentation room they visited the other night. Nevaeh is sitting at a table with another woman, each of them enjoying a beer. Cooper recognizes the stranger, the same one who called Nevaeh out of the room during the presentation: blonde hair with streaks of green, thin lips, taut skin. A small black container sits next to her.

The huge banner of Earth 2100 looms on the wall; Cooper's heart begins to beat faster. Just seeing this poster is enough for his threat response to react as if in the presence of a bear.

"Hello again, Pando," says Nevaeh as she stands, reaching out for a handshake.

"It's good to see you again so soon," he says.

"I'm glad you got our message," the leader notes. "And I see you brought your friend." She looks to the boy and shakes his hand. "It's Cooper, right?"

"Yeah," he says. "And you're...Nevaeh?"

She nods with a grin.

"That was quite the presentation last week," Cooper says.

"I'm glad you thought so."

Tierra looks at her sister and bows her head, then turns to exit the room.

"Thank you, Sister," Nevaeh says to her younger sibling.

"Yup," Tierra replies, heading back for the bar.

Nevaeh turns to the two recruits. "I'd like to introduce you both to my friend," she says, motioning toward the young woman now standing behind her.

"I'm Iri," the blonde girl says, reaching out for handshakes.

"Do you guys want a drink?" asks Nevaeh.

"Sure, I'll take a drink," says Pando. She nods and looks to Cooper.

"I'm good, thanks," he says.

Iri grabs a bottle out of a small fridge and pops it open on the side of a table. She hands it to Pando. "What did you guys think of the meeting on Thursday?"

"Awesome," says Pando.

"Terrifying," says Cooper, pointing his head in the direction of the heat map of Earth.

Nevaeh nods in agreement. "I think a little fear is healthy when it comes to climate change." She takes a seat at a table, the other three following her lead. "Now Pando, you mentioned that you were interested in learning more about what we do here at Brooklyn Project, is that correct?"

"Absolutely," he says. "It seems like you have a pretty good gig here."

The black woman takes a sip. "Well it just so happens that there's a project we've been working on that might interest you two," she begins. "Before we get into that, I want to learn a bit about you. What's your interest in climate change?"

Cooper and Pando glance at each other.

"Well, my family's hometown was leveled by a hurricane in the Philippines," Pando starts.

"You mean Kabubwason?" asks Iri. He nods.

"Oh my God, is everyone okay?"

Pando looks down and shakes his head. "But at least I'm still here, I guess."

"I'm really sorry to hear that, Pando," Nevaeh says. "You know we have support groups for people whose lives have been devastated by freak weather events," she reminds him. "What about you, Cooper?"

The boy searches the room awkwardly, wondering why he's been called here. "Uh, I dunno. It seems like climate change is a pretty big deal, especially for young people like me. I guess I just want to learn what I can do to help," he answers clumsily.

"You're right," says Iri. "It's the biggest of deals. I'm glad you're willing to fight for justice."

"Of course we are," says Pando.

"Do you guys think you can each dedicate some time to work with us?" asks Nevaeh.

"Depends," says Pando. "What is it exactly that you want us to do for you?"

Iri smiles at Nevaeh. "Take a look," she says, grabbing the black plastic box beside her and unclipping the top.

"As you know, Brooklyn Project is more than just an educational organization," the blonde explains. "We use a sizeable portion of our donations to fund clean energy development and scientific studies related to climate change." She pops off the lid, pulling out a device with a vague similarity to a transistor radio, but reflecting modern sensibilities and design trends, with two large round speakers and an unlit digital display.

"One of them involves monitoring shortwave radio frequencies to measure disruptions in the atmosphere." She places the radio on the table with a loud smack. "I could bore you with a bunch of science, but the details aren't important. Bottom line is, we want you guys to help us with this." She taps the radio device suggestively.

"Sooo...what is it you want us to do, exactly?" asks Pando.

She smiles. "This is a shortwave radio receiver." She depresses a button on its top, bringing the device to life. A low crackle is heard from the speakers. "We want you guys to listen to specific radio frequencies and let us know if you hear anything out of the ordinary."

Pando's heart sinks. This isn't the job he was looking for. "That's it?" he asks.

"That's it," she says encouragingly. "Simple. We'll give you the frequencies, you tune in, you come back to us and report what you hear. And then we pay you."

"You're gonna pay us?" asks Cooper.

"Ten dollars an hour, plus a bonus if you hear anything worth mentioning."

Pando's eyes widen with delight. "For real?" he asks. "So I could just have this thing on while I'm sitting in bed, and as long as I'm listening, you'll pay me?"

"You got it," says Iri.

"How will you know we're actually listening?" Cooper inquires.

"The radio has a built-in hard drive that records everything it receives when it's on. So every time you turn it on, it will be like punching in for work. It will start recording, and you'll be on the clock for as long as it records."

The gears turn in Pando's head. He's found a way out of unemployment.

"The device can only hold 12 hours of audio, which is why you'll have to come back here every two days to offload the data and report back what you've heard."

Pando is still impressed by the opportunity, and more than a little giddy. Sixty bucks a day to sit on his ass and listen to the radio. "Well I'm convinced," he says emphatically.

Cooper's face, meanwhile, is twisted into confusion.

"Glad to hear it," says Iri.

"So you're in?" Nevaeh asks them.

"Hell yeah," he answers unflinchingly.

Cooper shrugs off his pouting gut. "Yeah, I don't see why not," he grimaces.

"Excellent," says Iri. She flips off the receiver and replaces it into the plastic box. "Then this is for you. It's already been tuned to the frequency we want you to monitor. Just take it home, and when you have the chance to listen to it, just push the button on the top." She hands them the kit gingerly. "Now as I understand it, you're both underemployed, is that right?"

"Is there anyone our age who isn't?" cries Pando.

She smirks. "Then at least one of you should be able to report back here every two days at 5pm?"

They look at each other and nod. "Yeah, shouldn't be a problem."

"Good. We're counting on you guys." Iri rises to her feet, the rest of them following suit. She extends a hand. "See you in two days."

Pando grabs the shortwave off the table, followed by his beer glass.

"You can recycle your bottle in the front," says Nevaeh.

"Cool," he says, ready to leave. "So when do we get paid?"

"Two days."

"Word," Pando says. "See you then."

Cooper's still not keen on the specifics of their job. "So how will we know when we've heard something out of the ordinary?" he asks.

Iri looks to Nevaeh. "You'll know when you hear it," she says vaguely.

Cooper just stares at her.

"See you Wednesday at five," she says, giving them a curt nod.

Pando and Cooper depart through the curtain. The refugee bows his head to Tierra as they exit, "Thanks," and tosses his bottle into the blue bin. It lands with a clink.

As Cooper and Pando walk toward the train, they wonder about what they've just signed up for. Something about the whole thing doesn't sit well.

"I still don't know why they wanted me there," Cooper whines.

"You're so ungrateful," says Pando. "We just landed the easiest job in the world, and it's actually doing some good for it." They walk down the subway stairs in lock step.

Back at BKP, Iri looks to Nevaeh in the now-empty room. The well-worn routine again went as planned.

"You think they bought it?" she asks.

Nevaeh shrugs. "We'll see."

. . .

Shree slams the front door behind her, letting out an audible sigh. She finds Cooper and Pando sitting on opposite couches in the living room with their computers open. On the table between them sits the black shortwave device with an illuminated blue screen.

Pando flips his head upside down to see her. "Hey girl," he says.

She grunts unhappily. "I hate my job," she fumes as she climbs the ladder with a heavy bagful of tests to grade. She retreats to her room and slams the door shut.

Pando shoots Cooper a look of concern. He frowns sourly as he gets up to follow her. When he enters her room, he finds her body splayed out face-down on her bed. He takes a seat beside her chest.

"What's wrong?" he asks.

"I had such a shitty day," she says, her voice muffled by bedsheets.

"Tell me about it."

Shree lies there quietly. "The principal said she'll discipline us if our students aren't testing well," she discloses. "And you know how great my kids are at exams."

"That's fucking ridiculous," he says. "These people shouldn't be in charge of a school."

"Maybe," she speculates. "Or maybe I just shouldn't be a teacher."

Pando doesn't agree. "Oh stop it. You'll get their grades up."

"God willing," she whispers.

He looks down at her with agitation. "God," he begins, "has nothing to do with it. The genie in the sky isn't gonna teach those kids. This is about you. *You're* gonna do it."

Shree rises sharply in her bed. "God," she snaps, "has *everything* to do with it."

Pando frowns. "What you call God, I call futility," he says.

"What you call futility, I call destiny," she retorts.

"What you call destiny, I call disaster."

Shree is silent for a moment, his meaning sinking in. Her thoughts turn to Nevaeh's presentation, and with it, the Book of Revelation. Destiny. Disaster. Flooding. Flame. Apocalypse. The end days, baked in and fast approaching.

"On that," she murmurs, "we agree." She throws herself onto the mattress, burying her face in the pillow.

Pando brings his supple hand to hers, and she clasps it lovingly.

"My job kinda makes me want to kill herself," she admits.

Pando nods. "Climate change kinda makes me want to kill myself," he grumbles.

"You kinda make me want to kill myself," she jokes. She lets out a long groan, then exhales loudly. She doesn't stir.

"You'll be fine," he assures her.

No response.

Realizing she'd rather be left alone, Pando heads back downstairs. As soon as he leaves, Shree pulls out a Bible, and flips to a random page. She begins to read.

Downstairs, Cooper is reading about the *Titanic* disaster from his laptop. The unsinkable ship sank in less than three hours on that night in 1912. For a full day in advance of the disaster, several ice warnings had been received about the danger in their path. In hindsight, given the clear and present danger along their course, the people in command of that steamer should have exercised caution and changed their heading, or at the very least slowed down.

As it so happened, extreme warmth played a role in the mighty ship's demise — exceptional heat in the years before 1912 encouraged the melting of ice sheets in Greenland, causing icebergs to calve off like dominoes. One of them became the most infamous iceberg in history.

Pando makes his way down the ladder. "Anything on the radio?" he asks.

"Not a thing," Cooper answers.

The refugee sits across from him on the futon, popping open the beer that sat by its foot.

"I think Shree is depressed," Pando reveals.

"I think I'm depressed," Cooper echoes.

Pando frowns. "Yeah, well, me too." He takes a swig. "This is real life, man. Real life is depressing sometimes." He returns to his computer to browse his friend feed. The first thing he sees is a post from Shree:

> Shree Dunlap: The land must not be sold permanently, because the land is mine, and you are but aliens and my tenants. Throughout the country that

you hold as a possession, you must provide for the redemption of the land. Leviticus 25:23–24 5:53pm

Pando scoffs. "Ya know, I can't understand how a girl like Shree can believe in both science and religion at the same time," he says. "It's baffling."

A door squeaks open upstairs; Shree starts descending from the loft. "What are you losers up to?" she asks, joining Pando on the pink futon.

"Just 'working,'" says Cooper, pointing to the black device on the table. A low static rustles from its speakers.

Shree has brought a thin book and a small case of colored pencils. She flips open to one of the pages, revealing an image in outline form. She fumbles open the paper lid and selects a bright yellow color, pulling it out tenderly. She grips the rod with delicate fingers and begins shading in the picture.

Cooper studies her curiously. "What are *you* up to?" he asks.

"Coloring," she says. "It's relaxing."

"Rough day?" he asks her.

"You have no idea."

"People are being stupid," Pando explains.

"For real," says Shree. "Like, I can't even. I'm over it." Exasperation laces her voice.

Cooper winces. "Eh, it could be worse."

"How?" she asks, not looking up from her coloring book.

"Well at least you're, like, a respected staff member at an institution of learning," he points out. "At least what you do has value and meaning. At least you know you're making a difference. All I do for a living is take orders. There's no meaning in any of it."

The teacher shrugs. "I don't always feel respected," she says. "And I don't always think I'm making a difference."

Cooper feels deflated. Nothing he says ever cheers her up.

"But at least I'm trying," she adds. "I think we're all just trying."

He nods with a stiff lip, returning his gaze to the text on his screen. "So

I'm learning about the *Titanic*," he says, steering the subject to less personal matters. "There's a lot of crazy stuff that went into that ship."

No one acknowledges him; they're too engrossed in their own business. He goes on.

"Like, did you know the shipbuilders had to completely dismantle their gantries and rebuild them from the ground up? The old ones weren't big enough."

"Huh," says Pando, moderately fascinated. "You know what else is interesting?"

"What's that," Cooper asks.

"Some dude wrote a book about the *Titanic* before it even happened."

Cooper cocks his head in his direction. "What do you mean?"

"Like, some writer wrote a story about a ship sinking just like the *Titanic*," he explains. "He basically predicted it."

Cooper eyes him incredulously. "You're not serious," he says. "That's bullshit."

"No funny," Pando avers. "Look it up."

Cooper opens a new tab and types "writer predicted titanic." The results load instantly. He selects the first link, an encyclopedia entry entitled "*Futility, or the Wreck of the Titan*," by Morgan Robertson. He begins scanning the page.

"You're shitting me," he says.

"I know," Pando concurs. "Mind blown."

"Now I know what to put on my Christmas list," says Cooper. The coincidences between the novella and actual events are indeed astounding. "Even with fourteen years notice, they still couldn't avoid tragedy," rues the boy.

"Humans are dumb, Coop," Pando responds cynically. "Remember that."

Cooper looks at him critically. Over the radio, the static continues its uninterrupted droning.

Hammer to Fall

The day he knew was coming has finally arrived; Cooper sits atop the cold metal rafters at Columbia's winter graduation ceremony. He's there to support Natasha, hoping he'll be able to meet up with her afterward. He watches as the graduates, no different than him, walk one by one across the indoor stage. Step, step. Tick, tick.

Natasha is just another name on the list in the stuffy auditorium. There she goes, $200,000 paper in hand, traipsing cross the stage with confidence. And just like that, she's walking back to her seat again. Four years of work led up to that ten-second promenade.

Despite the applause, the affair feels like a funeral to Cooper: the death of the promises of the future. It all feels so hollow. These people are clapping, yet the planet is dying. To make matters worse, now that Natasha's graduated, he knows he won't even see her anymore. He's too busy to attend her environmental club, and in a city like this, no one has time for anything unless it's in their own borough. But Natasha lives in the Bronx, unacceptably far away. She may as well live on the moon.

A buzz rings out from his pocket.

Pando: hey bro i totally forgot im hanging out with shree later. can u take the radio to bkp? 2:30pm
Cooper: 😧 2:30pm
Pando: thanks man 2:32pm

Dick move, Cooper thinks. He sees Natasha has reseated herself. The rest of it doesn't matter; his mind is left to wander. He fantasizes about getting together with her, the girl he's been fixated on since the day she raised her hand in class.

The ceremony ends. He works his way through the crowd, zeroing in on where Natasha was last seen. After a brief search, he finds her in her sky blue gown, all smiles. He gives her a loving hug.

"You're all grown up now," he says to her proudly.

"I know, I can't even believe it," she concedes.

"Now go change the world," he mandates.

The girl laughs charmingly. "I'll do what I can," she promises. "Are you ever gonna come to a divestment meeting?"

The boy looks at his feet. "I really want to, but I've been so busy with work."

Natasha understands. "Well just make sure you come to the protest in May," she tells him. "I'll still be on campus every now and then, so you'll probably see me around."

She'll still be around. Thank God. His mood brightens considerably. "Yeah, well come get coffee sometime," he says. "I'll give you a discount."

"Aww, you're so sweet," she says, rubbing his shoulder gently. "I'll let you know."

Just then someone glomps her in surprise, a friend Natasha recognizes immediately, elation on her face. Within seconds, she's swept away by the collective motion of the crowd as they all make their way for the exit.

And suddenly Cooper feels completely alone.

. . .

With black radio in hand, Cooper has returned to the BKP presentation room with Iri, who's quite happy to see him.

"How's life, Coop?" she asks.

The boy hates hearing that question. Can life ever be good when the world is collapsing? "Fine," he answers.

Iri looks at him with sea green eyes. A brown flower bandana contains her dyed blonde hair, wispy locks of yellow and blue draped over it. She has the poise of a mature woman, but the grit of a spirited warrior. That she's out for blood isn't immediately apparent to young Cooper.

"How old are you again?" she asks him.

"21."

"A good year," she muses as she opens a clamshell laptop. Cooper takes a seat on the opposite bench, placing the shortwave on the wooden tabletop.

"Did you hear the news?" she asks. "November was the warmest on record. This will probably be the hottest year in the books."

Cooper swallows uncomfortably. The march to disaster continues unbroken.

Iri drags the shortwave beside her computer. She pushes the device's power button three times, wirelessly pairing the radio to her laptop. She begins looking through the data.

"I see you guys maxed out your hard drive space," she notes. "Anything interesting?"

"Not a thing," Cooper reports. "Then again, I don't really know what I'm looking for," he adds.

She hmms. "Well I'm recalibrating the frequency for you, so this time might be different."

Cooper sits in silence as she types loudly on her keyboard.

"So what have you been up to lately?" she asks offhand. He finds the unsolicited small talk tenuous and uncomfortable. He lets the silence play out.

"Have you been reading through the information our website?" she asks him directly.

"More or less," he answers. "I read the page about El Niño. It's fascinating stuff."

"You're damn right it is. I find the part about the tornadoes particularly frightening," she says.

"Yeah," he agrees, his chest growing tight at the thought of it. "I lost my mother to a tornado."

Iri looks up from her laptop screen. "I'm so sorry to hear that," she says sincerely. "It must have been hard growing up without your mother."

"It is," he replies. "Feels like a lot of people are losing family to crazy weather shit these days."

She agrees with a nod. "Get used to it, my fellow climate warrior." The radio goes blank as it disconnects from the laptop.

"But that's why we have people like you," she reminds him, handing back the device. "Here you are, ready for another round of listening," she says smiling.

He takes it from her. There's a question on his mind.

"What is it?" she asks.

He wavers. Sometimes it takes all the courage in the world to ask a simple question.

"Do you ever wonder what the point is?" he asks. "Like, if the world as we know it is falling apart, then what's the point?"

Iri considers him. "The point is to not let that happen," she proposes. "No matter what it takes, we won't let that happen."

Cooper nods, understanding. If the world is ending, then the point is to prevent it. Otherwise, none of this really means anything, does it?

"Oh, before I forget," she says, reaching into her pocket to pull out a number of folded green banknotes. "For your hard work. You can split it with Pando however you two decide."

He reaches for them sluggishly. "Thanks," he whispers, considering the currency in his hands. He shakes his head with uncertainty. "Ya know, Iri, I'll be honest with you, this whole thing seems mad sketch."

She looks at him defensively. "What do you mean?"

"I mean, I just don't get this," he says. "I don't get what we're supposed to be doing here."

She thinks about how to respond. They can't afford to lose someone like him.

"Let me ask you this. Are you afraid of global warming?"

He blinks. "Uh, yeah."

"Okay. Do you believe in the Brooklyn Project and what we stand for?"

He glances over at 22nd century Earth. Its blood-soaked face rages at him.

"Sure, yeah," he says.

"Do you want to be an alarm clock and wake up this sleepwalking world before we fall off the cliff?" she asks pointedly.

"Of course I do."

She picks up the radio and puts it in front of him. "Then do your job, and I'll see you in two days."

His eyes narrow. There's a glimmer in her expression that he can't read, but he trusts her nonetheless. He picks up the device.

"Will do," he agrees. He gets up from his bench and turns to leave.

"See you Friday," she says. Cooper nods uncertainly, and then disappears behind two velvet lips. Alone, Iri walks to the edge of the room and grabs a clipboard off the wall. She puts a checkmark next to Cooper's name, right below the check next to Pando's.

. . .

The radio crackles in the living room; Pando is barely listening. Shree has a white bud in her ear, her head bobbing as she browses her phone. Mel walks in, frustrated.

"How was your date?" Shree asks her.

"It was so meh," Mel says, taking a seat opposite the couple. She hears the radio stirring unsettlingly on the table. It's the only noise in the room. Unhappy with the relative silence, she opens the music app on her phone and selects a stream. Hip hop plays from the tiny speaker.

Pando glowers at her. "Do you mind, Mel?" he growls, pointing to the radio with his eyes.

She sighs. "You're still listening to this ridiculous thing?" she asks.

"It's my job, actually," he reminds her.

She scoffs and pulls out a pair of earbuds from her purse. "Why don't you get a real job," she says, plugging her ears. The music from her phone goes mute, rerouted to the white speakers.

"Why don't you quit being a bitch," Pando utters under his breath, returning his focus to the computer screen in his lap. Mel's music is too loud for her to hear.

Cooper, sitting in his room, heeds the call of hunger. He descends the ladder to enjoy his daily meal, which today will be eggs. Reaching for the handle of the fridge, he pauses before he opens it, noticing the enormous pile of dishes in the sink. One of them is the lone skillet. He pulls it out from the stack and turns on the water, but the plates are up to the faucet; there isn't any room to wash it properly. Frustrated, he wedges it back into the sink.

"Guys, we have a mountain of dishes to wash," he tells them.

"Oh you noticed that, huh," says Mel.

"Why aren't we doing dishes?"

"Most of them aren't mine," she maintains, leering at Pando.

"Literally one plate in there is mine," the refugee says angrily.

"Well I eat like once a week, so they sure as hell aren't mine," Cooper says.

Shree is too focused on her phone to care, and Pando has returned to his laptop screen. At an impasse, Cooper returns to the sink and turns the faucet on full stream. He grabs the sponge and the container of dish soap, only to find there isn't any left.

"Who's buying soap this week?" he asks, raising his voice.

"I bought last week," says Shree.

"I got the new sponges," says Pando.

Cooper is frustrated by the lack of initiative. "Alright fuck it, I'll just go get some," he says.

Mel claps at the idea. "*Thank* you, Cooper. It's nice to see someone stepping up to clean," she applauds. "If you get the soap, I'll get the dishes."

He grabs his keys and wallet from his room and throws on a sweater, leaving the apartment with a scowl on his face. A smudge of dried vomit sits on the sidewalk beside the stoop; Cooper walks by without interest. He's more concerned with his personal safety. His body is painfully vulnerable; his muscles have atrophied from famine. Now he walks alone on these open season streets, where predator and prey keep watchful awareness at all times. He knows too well on which side of the divide he falls.

He turns into the brightly lit bodega at the end of the block. There are only two aisles; he finds the dish soap quickly. There's only one variety to choose from, a brand name that probably costs more than it should. With no alternatives, he grabs it and joins the line for the register.

"No, not the big one," says the woman in front, haggling with the cashier. "I don't need that much."

"Not the big one," the cashier replies in a foreign accent.

"How much the big one cost?" the woman asks.

"Two dollars."

She considers it for a moment. "Okay, I'll take the big one."

The register opens; money is exchanged. The woman leaves, allowing Cooper to advance. He puts the dish soap on the counter.

"3.50," the cashier says.

Yikes, Cooper thinks, not realizing it would be so expensive. $3.50? That's like half a pizza. But it's already sitting on the counter; he should just buy it. It'll wash the dishes for a month, or maybe a year at the rate they wash dishes.

He reaches into his wallet for his credit card. He hands the plastic to the cashier.

"Five dollar minimum," the man croaks.

"Oh," Cooper says, withdrawing his hand. He fumbles through his cash pocket. Three dollars. Shit. He'll have to buy something else.

"How much is a candy bar?" he asks.

"Which one?"

Cooper points to a sleeve of bright orange packaging.

"Two dollars," the man says.

Okay, Cooper thinks. $5.50. He gives him his card and walks out with his goods swinging from a white bag.

Today I've added one plastic bag to the world, he thinks, his heart sitting heavy in his chest. He tries to put it out of his head, but he can't. This poor, sick world. This gravely ill planet. It's hardly a case of it being incurable, but rather the way it feels like it's already dead. Who wants to believe that our way of life could be such a death sentence?

In the distance, he can see the silhouettes of skyscrapers across the East River. They stand out like super-sized versions of the stone moai of Easter Island, towering symbols of human society. But instead of looking powerful and eternal, they appear transient and fragile. No longer does the city look like a gleaming metropolis to him, but rather like giant stone skulls prodding into the sky. These are buildings of bone, to be dug up by the future archeologists of the world, if there are to be any future archeologists. The skyline is less a vibrant capital of the world filled with life and hot blood, and more a harbinger of its own downfall, its final anemic years playing out in stubborn denial before the long darkness takes hold. These buildings, Cooper realizes, are mere pillars of oppression, a last glimpse of the prosperity that was, before humanity enters the bitter ending we've condemned ourselves to die through. He can't shake the feeling that he's just another dead citizen of a dead civilization, a society as fossilized as the fuel that powers it.

Cooper returns to his apartment. Nobody has moved from their seats; they're all transfixed by their glowing devices. Mel is playing a colorful game on her phone. Shree is browsing her friend stream. Pando appears to be reading an article. Minus the crackling radio, there is no noise.

"Here you go, Mel," he says, putting the soap bottle on the counter by the sink.

"Okay," she says, "as soon as I beat this level." She's blasting an onslaught of gray aliens as she flicks her fingertips across the screen. Cooper watches as a ferocious group of extraterrestrials bum rush her screen. Her fingers don't move fast enough to mow them down. They pounce on her character's avatar, dragging her down violently. The screen fades white. Game Over.

"Fuck," she says. "Stupid game." She heads to the kitchen.

Cooper takes her seat on the futon, and he pulls out the partially-melted candy bar in his pocket. He peels back the crinkly wrapper and sticks the pleasure directly in his mouth.

Pando looks over to him. "Nice to see you're investing in your health," he says sarcastically.

"Hey, life is short," Cooper says. "Eat dessert first."

Pando chuckles. "Here, I have something that might make it even better." He pulls out a box from under the couch and opens it. Inside is his drug paraphernalia: bubbler, lighter, grinder, and a dirty bag of bud. He hands it to the boy.

"Thanks," Cooper says. He grinds up a nug and packs the filings into the bowl. With lighter in hand, his thumb snaps the flame alive, and he draws it through the green psychoactive with a deep breath. He inhales the white smoke — the vaporized cannabinoids enter his lungs, where they're picked up by the cells in his blood as they pass over the alveoli at the terminus of his respiratory tree. The blood cells, impregnated with the wondrous chemicals, continue their perpetual race through a hundred thousand kilometers of arteries and veins that twist through Cooper's body, bringing life-saving oxygen to every cell. Eventually they reach his brain, where the cannabinoids bind to receptors on his neurons, making Cooper fantastically high. All within the space of a minute.

He hands off the pipe and returns to his bar of chocolate. Pando medicates with a large hit, finding himself instantly baked. The nerve cells in his intoxicated brain begin to fire rapidly, his thoughts racing. He looks back at the news article on his screen, and his face curls into an expression of disgust.

"What are you reading, Pando?" Cooper asks, noting his dismay — or is it fear?

"Oh, just the news," he grumbles. "Our do-nothing Congress failed to pass another carbon bill. Idiots." His uninhibited mind takes him down dangerous roads of thought. "The world would be so much better off if it was in the hands of one smart human instead of a handful of dumb ones," he thinks aloud.

Cooper looks at him with concern. "Oh yeah? And who would that one smart human be?"

"Me, preferably," he says. A smile creeps over his face; they both break into laughter.

Mel overhears this exchange from the kitchen, the sink running quietly. "Pando, I'm pretty sure if you were in charge of the world, I would kill myself," she says.

"Maybe you should anyway," he jokes.

"You're right. Living with you is bad enough."

Pando smirks. "You may be wrong about a lot of shit, but I like your spunk, Mel," he admits. He stares back at the screen. "I think we should take all the deniers and put them on an island somewhere, or maybe just kill them all, before they kill us."

"I'm a denier," Mel interjects, proudly flaunting her ignorance. "You really want to kill me?"

"You?" he responds with a smug laugh. "More than any of them."

"Even after all the dishes I do for you?" she teases.

"Anyone who doesn't believe in climate change is complicit with mass murder," he sounds. "It doesn't matter who you are. If you deny the facts, you're wrong."

"You're right, Pando, I'm a mass murderer," she says.

"Quit being mean, you two," Shree says disapprovingly.

But to Pando, he isn't being mean; he's just telling it like it is. The world to him is nothing but the dichotomy between true and false, 1 and

0, and there is no in between. If you aren't right, you're wrong. The distinction is clear. And he knows when he's right.

"I just don't get it, Mel, how you can be so thick," he starts again. "You *saw* the data. You saw the graphs. What more convincing is there to do? Are you blind?"

Shree sighs, helpless to stop the verbal sparring.

A shriek peals out from the kitchen. It's Mel.

"What is it?" asks Shree.

"Oh my god, I just saw the biggest cockroach ever!" the blonde shouts, dropping a plate into the dishwater. She steps back from the counter hurriedly.

"Where?" asks Pando, running toward the kitchen. Shree follows behind.

"Right there!" Mel screams.

Pando sees a huge creature on the countertop staring at him, its antennae flailing rapidly. "Holy *shit,*" he says.

Shree hides behind him to see the monster. It's the size of a fist.

"Give me your flip flop," Pando instructs. Shree obeys immediately, pulling off her sandal for him. He moves slowly toward the arthropod, careful not to startle it.

Sensing his approach, the bug panics, tearing off in a frenzied sprint. Pando's mind, under the influence of the psychoactive, tracks his target's movements in slow motion. He brings the worn shoe down on its scurrying body with all the power his strength commands. *Thwap.*

Mel and Shree hold their breath. Cooper watches from the futon.

Pando slowly lifts the yellow sandal from the impact site, revealing the crushed body of his victim. There isn't even a twitch to indicate any remaining life; it has been deftly extinguished.

"Jesus. Good job, Pando," Shree says.

He grabs the bug by its muscular leg and brings it to the trash. Relieved, Pando flings its body into the garbage. "You're welcome," he says confidently.

Mel peers into the wastebasket with a smile. "You see, Pando?" she says. "You're a cold-blooded killer like the rest of us."

"Oh please," he responds. "Last I checked, cockroaches aren't an endangered species."

"Uh-huh. Sure." Mel resumes the last of the dishes as the rest of them return to the living room.

Cooper returns to browsing ties on a shopping app from his phone. *What to get Dad for Christmas?* he ponders. *Dad never wants anything. Dads are so hard to please.*

Mel reaches the bottom of the dish pile, but before she drains the sink, she notices a small collection of white plastic forks floating in the soapy water. She picks them up curiously.

"What're these?" she asks, holding the dripping plastic over her head for her roommates to see.

Pando looks over. "They're forks," he says astutely.

"They're disposable. Why are they in the sink?"

"Because they haven't outlived their usefulness," he answers.

She pulls the drain plug and walks over to the wastebasket. "We have enough forks," she says, throwing them into the trash.

"Don't throw those away," he urges.

"I just did."

"Then take them out."

"Ew, no," Mel protests. "There's a dead cockroach in there." The sink coughs as the water finishes draining. "So are you guys gonna just sit here in silence?" she asks. No reply. The shortwave continues to crackle.

She decides to head to her loft. Once inside, she closes the door and starts blasting music from her computer. It's just loud enough to be a nuisance to Pando, but not loud enough that it's worth a confrontation.

"Can you believe her?" he asks Shree, rising to grab the forks out of the trash. He returns them to the sink, from which a stream of water is still trickling from the faucet. "And she left the water running. Who the hell does she think she is." He rinses off the forks and puts them in the dish drain, ready to use them again. Like Cooper, he lives in a state of poverty, where conservation is the difference between life and death. Everything is used until it can't be. Even plastic forks.

He sits back down. Cooper is now lying on his back, pawing the air aimlessly, clearly a bit higher than normal. The silence continues.

Then, out of nowhere, a noise blares from over the radio. It startles Cooper upright.

"The fuck?" he groans. It sounds like music played on a piano, but like no piano he's ever heard. It sounds possessed, haunted, distant and distorted. It sounds like a ragtime nightmare.

Shree puts her phone down to listen; Pando's eyes are locked on the radio, the disturbing melody playing over the speakers.

"You think this is it?" asks Cooper.

"Shh. Listen." Pando quickly opens a new file on his computer. He notes the current time: 8:45pm.

The music stops abruptly, replaced by the disquieting voice of a little girl speaking a line of numbers. "Five. Seven. Two. Zero. Eight..." it begins, in the staccato manner of a speech synthesizer.

Cooper looks at Pando wide-eyed. The man begins to write.

"Nine. Four. Four. Seven. Eight. Three. One. Five. One." The chilling voice continues in its unnatural rhythm, like the sound of a demon possessing a toddler. "Six. Six. Zero. Zero. Five. Nine. Seven. Nine."

Pando writes each number as it's said. "One. Zero. Eight. Six."

Then, just as suddenly as it starts, it stops, cutting off after 44 digits. The crackle of the static returns.

Pando looks to Cooper, who's staring back at him, mouth agape. Shree's eyes flit back and forth between both of them.

"What the hell was that?" she asks.

The men stare at each other.

"I have no idea," Pando says. "But I think it was important."

Invisible Sun

Cooper and Pando enter the backroom of Lucky Dragon No. 7, shortwave in hand. Iri is sitting at a table with her laptop open, reviewing the latest data from the ongoing war.

"We heard something," Pando announces as he walks in.

She looks up casually. "What did you hear?" she asks.

He approaches her table and places the radio beside her. "A really creepy broadcast," he describes. "It started with a piano tune and then there was a little girl speaking numbers." He hands her a slip of paper with the 44 digits on it.

Iri looks at the message in silence, not immediately sure what to make of it. Then her eyes go dark. She snaps her fingers repeatedly, like she's trying to remember something. Unable to recall, she gets up to walk over to a filing cabinet in the corner of the room.

Her thumbs dance over manila folders. She pulls one out, revealing a piece of paper inside. The words "coke # station" are scribbled at the top in red ink. With Pando's paper in hand, she compares notes.

"You found it," she gasps.

"What is it?" Cooper asks, stepping forward.

"Coke."

The word comes forth with trepidation, a cursed name with evil connotations. Cooper looks at Pando with confusion.

Iri replaces the paper back into the folder and files it away. With a deep breath, she looks at the two young men distressingly. She waves them to follow her, leading them through a side door to a briefing room. She locks the bolt behind her.

"Do not breathe a word of this to anyone," she tells them.

They take a seat at the table nervously, perspiration beading on Cooper's forehead. He watches as Iri shuffles over to another huge file cabinet, one of four in the room. She grabs a thick collection of paper.

"Senator Joseph Coke," she says, pointing to the inset picture of the wrinkled man on the top of the stack. "75-year-old climate change denier. Chairman of the Senate Science Committee. Son of oil baron Frederick Coke. And traitor to the human race."

Cooper looks at the geriatric man smiling in the picture. "That's the guy you were telling us about in your presentation," he notes.

"All deniers are traitors," Pando affirms.

"But he's the worst of the worst," Iri says. "The coldest blood of them all."

Goosebumps snake along Cooper's skin, accompanied by a shudder down his spine. "What makes him so bad?" he asks.

"Because he runs the Coke Coalition," Iri explains, "a group of carbon junkies that fund climate change denial. For decades, they've been sowing the seeds of doubt and indecision that have led us to where we are now. They're the reason we're in such dire straits."

"Carbon junkies?" Cooper asks. "Like carbon polluters?"

"Well, we're all carbon polluters," Iri reminds him. "But yes, these are the true junkies, the ones that won't even acknowledge our addiction. These are the people that dig carbon out of the ground for profit."

She flips the page, revealing a tactical map of the United States divided into greens and reds. To the north and west are pockets of green: New York City, Chicago, Philadelphia, Los Angeles. To the south, a wall of red.

"They're the reds on this map," she explains. "And we're at war with them."

Stress hormones fill Cooper's brain, traveling into his blood and throughout his body. "So there's a lot more to the Brooklyn Project than we've been led to believe," he says.

"You're very observant," Iri says. "Education alone isn't going to save the world. We have to stand up and fight."

Pando finds the idea irresistibly enticing; they aren't all talk after all. He studies the map. "So you're the greens, I take it."

Iri smiles at him. "The Clean Front, united in a common purpose — to decarbonize America." She points to the red line on the map. "The reds don't have the sheer numbers that we do, but they have way more money — the entire fossil fuel industry is on their side. Billions of dollars piped right into their wallets."

"Why do you call them reds?" Cooper asks.

"Because of their ties," she answers. "All deniers fall on one side of the political spectrum."

She flips to another page. "We've learned that the Coke Coalition is secretly planning something big, something deadly." She gestures to a diagram on the page, a schematic of the geoengineering device she had the unfortunate privilege of witnessing in person. "A geobomb."

Cooper shuffles in his seat with discomfort; he's almost afraid to know what it is. But he chooses knowledge over denial: "What's a geobomb?" he asks.

"A geoengineering device," she murmurs. "Or, perhaps more accurately, a doomsday device."

"Geoengineering?" asks Cooper. "What's that?"

"A solution, of sorts," she says.

"To what?"

"Climate change."

Pando scoffs. "More like a solution to life," he says. "It isn't a solution at all."

"I'm confused," Cooper admits.

"Geoengineering," Iri begins. "The junkies think the most cost-effective way to avoid catastrophic warming isn't by decarbonizing the economy, but by engineering the climate itself. The Cokeheads are building a device that would do just that." She points to the cylinders in the diagram. "These containers of aerosol are sitting in a warehouse in Mexico. The Cokeheads are planning to release this aerosol into the atmosphere, as a way to, in theory, reflect incoming sunlight back into space and reduce the effects of global warming."

Cooper rubs his chin, digesting this new idea. "So why is that bad?" he asks. "That actually sounds like a really clever solution."

"Hell no," Pando protests. "Are you crazy?"

"But wouldn't that solve our problem?" the boy asks.

Iri shakes her head. "Climate change isn't a heat problem. It's a carbon problem. If the Cokeheads deploy their weapon, it might lower the temperatures for a few years, but it would do nothing to solve the underlying issue. It would just give the world an excuse to burn more carbon, which is probably their plan."

Cooper leans back in his chair. Climate solutions are never as easy as they sound.

"As long as we burn carbon, the oceans will continue to acidify," Pando explains. "The food chain will still collapse. And that isn't a solution."

"These are evil men, Cooper," Iri says. "They're forcing us to eat the forbidden fruit at the point of a gun."

"They're evil?" Cooper asks pointedly. "This Coke guy doesn't look so evil to me. Clearly he has good intentions, if he's trying to stop global warming."

Iri frowns. "Intent isn't the issue," she says. "Even Hitler thought he was making the world better. But it doesn't matter what he thought, because he was wrong." She looks Cooper in the eye. "But these men aren't only evil because they're wrong. They're evil because they're about to wipe out 90% of life on Earth."

Mass extinction, Cooper reminds himself. Those are the stakes. It's all or nothing. Tick, tick.

"They're like supervillains," she continues. "Anyone who tries to destroy the Earth is evil, and they should be treated as such. Which is why we formed the Clean Front to stop them."

"So how *do* we stop them?" Pando asks, captivated by the militant rhetoric.

"You already are; you found their numbers station."

Cooper's eyes flicker with understanding. "So that's why you gave us the radio," he realizes. "You weren't performing some stupid experiment. You were looking for a broadcast."

More nodding. "That, and we had to judge your loyalty to the cause. You both passed, hence why you're here right now."

"What is a numbers station anyway?" Cooper inquires.

She rustles in her seat. "It's a way of sending messages without being tracked. Only the right cipher can decode the numbers."

"Why do the Cokeheads have a numbers station?" asks Pando.

"They have agents in the field," Iri reveals. "We don't know what they're doing. They could be anywhere. Finding their numbers station is one step closer to finding out what they're doing with it."

They could be anywhere, Cooper thinks.

"We installed a surveillance package in their warehouse," she continues, "but they discovered it within days. We don't know how far along they've come since then. We're in the early stages of planning another mission to the facility, maybe even to destroy it."

"Why not just go to the police?" asks Cooper.

"Because they aren't doing anything illegal, technically," she explains. "Besides, the junkies hold power at all levels of government. Most of them would celebrate a geobomb."

Cooper frowns at her. "I don't see why you're willing to risk everything you've done just to blow up a building," he demurs.

"Because engineering the climate is an unacceptable risk," she presses. "We have absolutely no way of knowing what the consequences would be if they release that much aerosol into the atmosphere. If their calculations are off by even the slightest amount, it could knock us back to the ice age." She brings her fist down over the diagram. "Do you really wanna take the one in a million chance that they *don't* destroy the world?"

Passion brims through Pando's body. A real opportunity to do something. It's finally here. "So what is it you want us to do?" he asks her keenly.

She considers both of them carefully. "We need you to join us to stop Coke," she says. "It's the only way to avoid the climate wars."

"The climate wars?" Cooper asks tenderly.

She nods grimly. "They'll be long, and they'll be bitter, and not even the United States of America will be able to afford them," she says ominously. "Every minute we wait is a minute closer to war drums. They'll bring civilization to its knees. And history will remember every single person who fought in them. Who lied, and who told the truth. Who tried to hurt, and who tried to help. And that's why we need you now, because we're losing this war."

"Sign me up," says Pando unconditionally. "This is too important to sit out."

Iri smiles into his eyes, which flash like mica in the sun. She turns to Cooper.

The boy is taking it all in. On some level, he knows they're right, but he doesn't want to accept it.

"I don't think I'm ready for this," he frets. "I still have a future to look forward to."

"Some future," Iri scowls, pointing again to the geobomb on the page. "You have to start from the assumption that the future is already lost, because right now it is. And the only way to get it back is to fight for it."

This resonates with Cooper. A people ought to fight for their right to exist. So should a species. The real question for him is, do we deserve that right?

"There was a time when our country split in two," Iri recalls. "When half of the country accused the other half of destroying its economy, destroying its way of life. The South said they'd never survive without their slaves." She flips back to the map in its reds and greens. "Well today, the reds are the slave masters, and we and our children are their slaves. And they'll be enslaving us to a hundred thousand years of irrevocable hell, living in chains, sacrificing our freedom for theirs."

Each word lands powerful blows on the boy's psyche. Still he stares at her, arms folded, not budging. But he feels the levees crumbling.

Pando steps in to help. "Coop, if I told you there are people trying to take your home from you, your country from you, would you just sit and let them take it, or would you take a stand?" he asks. "Because that's where we are."

Iri nods vigorously. "This is a revolution, and we need you."

Cooper soaks it up. "But what if the revolution fails?" he asks her.

"If it fails?" She looks deep into his eyes. "Then the human race dies."

He looks back at the file on the table. There's a lot of red on that map.

"So what do you say, Coop?" she asks. "Do you want to let the future wash over you, never trying to stem the rising tide? Or do you want to get in this fight, do something that matters? Now's the time to decide. Our generation is the only one that can stop the carbon junkies. There's no one coming to our rescue. It's on us."

His generation. Millennials. The people alive today. Last he checked, he's still alive.

Cooper sighs. "Alright," he relents. "What's my job?"

Iri pulls out a piece of paper from the back of the Coke file. It features a key, numbers to letters. It appears to be a code-breaking sheet.

"We acquired this from their database before they removed our surveillance pwnie," she goes on. "We think it might be the cipher to their numbers station." She hands the page to Pando.

"You want us to decode their broadcasts?" Pando asks.

"And investigate, if necessary," she adds.

"Investigate?" asks Cooper.

"The Cokeheads could be anywhere," she reminds them. "Our agents are stretched thin right now, and we have no one to pull if we need to check something out quickly." She flips closed the file and holds it in her hands. "Can you do this?"

The men look at each other and nod in agreement.

"What happens if something goes wrong?" Cooper asks.

"If you feel like you're in danger, leave," she suggests.

Cooper rubs his arm. "When do I *not* feel like I'm in danger?"

She smirks slightly before rising for the file cabinet. "You should have everything you need," she says. She turns around to face them, closing the cabinet with a bang. "Go save the world."

Pando folds the cipher into his pocket, and he and Cooper start for the door.

"Wait," says Pando. "Shouldn't we take your number or something?" he asks.

"I don't think I'm ready to give you my number," she says, smiling coyly. "We don't use digital communication unless it's absolutely necessary. They'd be able to track our every move if we did. Hence all the paper."

Satisfied, Pando departs confidently. Cooper follows, but looks back before he leaves. He stares at Iri with uncertain eyes.

"See you Wednesday at five," she charges.

. . .

"Eight, two, eight, two, five, zero, nine, zero," the spooky voice says in its metronomic tone. Cooper and Pando have moved their shortwave to Cooper's bedroom; they don't want their roommates asking any questions. If everything Iri has said is true, no one can know about their mission.

Pando fiddles with the cardstock cipher entrusted to him. As each number is fired off, he inputs it into the pad, producing a string of letters that he writes down one by one. The numbers come to an end; the distorted bars of the Maple Leaf Rag play briefly at the conclusion of the broadcast, and static returns to the radio.

Pando continues to scribble a translation. Cooper watches with baited breath, eager to know what it portends. After a minute of writing, he appears to have arrived at an answer.

"Well?" Cooper asks impatiently.

Pando's brow furrows. "I think it's some kind of signpost," he says.

"What do you mean?"

Pando hands him the penciled transcription. It's as enigmatic as the numbers themselves: "city hall station 4 am goods arrived."

Cooper looks up, bewildered. "What on Earth does that mean?" he asks.

His friend shakes his head absently. "If I even translated it right," he offers.

"Like the City Hall subway stop?" Cooper asks.

The gears turn in Pando's mind. "Maybe," he says. He scratches his head, searching for more insight.

"So they *are* here," Cooper says, his face flushing as he considers the implications of Cokeheads in New York. He can feel his heart pounding out of his chest; whatever they've found, it's important.

Pando stares hard at the transcription. "This could be it, Coop," he says. A shipment of goods, designed for who-knows-what nefarious purpose, coming tonight. "We have to go," he declares.

"But how? City Hall station is abandoned," Cooper points out. "There's no way to get to it."

Pando shakes his head. "There is one way," he warns.

Cooper looks around cautiously. "Dude. No fucking way."

"We have to," he says again. "Iri told us to investigate."

The boy says nothing. Pando's silver eyes persist.

Cooper frowns and grabs his computer off the floor. He punches in a query and reads the first result.

"The closest stop is Brooklyn Bridge," Cooper explains.

Pando nods assuringly. "We'll time our traverse to avoid any trains," he says. "They come like twice an hour at 4am."

Cooper glares at him. "I'm not fucking walking the tracks," he says adamantly. "That's suicide."

Pando's stare challenges his. "Oh, sack up, man!" he cries. "We have to go. You know what's at stake."

"But what if we get ambushed by Cokeheads or something?" Cooper shouts.

The refugee sighs quietly. "If there's anything dangerous," he says, "then we'll leave. But you know we can't just ignore this." He puts his hand on the boy's shoulder, tactile contact to persuade him. "I can't stop climate change on my own. And I can't stop the Cokeheads without you."

"So we're just gonna run down the tracks and hope nobody sees?" Cooper asks.

"If anyone catches us, we'll say we're just a couple of drunk kids looking for adventure," he offers.

"This is ludicrous," Cooper fumes. He marches angrily to his duffel, and he sifts through its contents for the one weapon he brought with him to New York, just in case he might need it — a red pocketknife.

Pando smiles with approval. "Do you even know how to use that thing?"

Cooper shakes his head. "But it might come in handy," he says.

His elder holds out his hand, and Cooper places it safely in his custody. "I'll take care of it," Pando says. "We leave in an hour."

The darkness of December envelops the cold New York night as the roommates set out on their first mission. They round the block and pass graffiti alley, a stretch of brick along their route covered with spray-painted tags in various hues of purple, black, and red. The glyphs look grotesque to Cooper's eyes, their meaning almost indiscernible in the dead of the night. The streetlights along this sidewalk have long since gone dark, and no one seems to have any interest in replacing the bulbs. There is no moon overhead; the few distant starts that poke through the city haze are all that light their way.

Cooper notes a tree planted along the street that has grown wider than its concrete base, its roots sending the sidewalk upward in great slabs. Beside it, a grease-stained bag sits on withered grass, lazily discarded by some human after stuffing fast food into its face hole.

The Brooklyn Bridge station is chilly and damp, completely deserted at this hour. The two young adults descend the stairs with strained nonchalance, trying not to draw attention to themselves as they take seats on the desolate platform.

The squalor of the subway is apparent: the peeling paint on the ceiling, the century-old signaling system, the garbage that litters the tracks. The structural poles are covered in a film of brown dust, faded yellow stickers clinging to their rusted backs. Insulation has been sliced open, decades-old duct tape peeling away at the edges. These are the arteries of this great city, but they're filthier than a landfill, and treated with the same level of dignity. New York is fucking old.

Cooper taps his foot nervously, trying to reduce his burgeoning anxiety. Pando sits stoically to his left, red scar ridged along his cheek. It looks hot with blood, an eternal reminder of what he's already been through.

"So how does it feel living in the adult world," Pando asks calmly, trying to break the tension.

Cooper responds with an almost sinister laugh. "I feel like an idiot," he answers. "My GPA has really helped so far."

As he looks around restlessly in the darkly lit tunnel, he remembers the naivety he approached the world with just a few months earlier, how easy he thought all this would be. After all, his grades told him he was smart; how hard could life really be? A lot harder than expected, apparently.

He wonders how long he'll consider New York his home. He recalls the television coverage of October 29, 2012, when the great city was brought to its knees by Hurricane Sandy. Visions of the fearsome storm whipping through the streets command his thoughts, and suddenly he finds himself lost in a vision of climate change's frightful wrath. He may not have been present for that hellish day, but he and the rest of America will never forget the terror it unleashed.

By the time it struck the East Coast, it wasn't even a hurricane, yet the crippling storm surge proved no less deadly. Waves crashed over the Battery and into Lower Manhattan, violently inundating the streets. The East River rose briskly, a catastrophe of epic proportions, the water sloshing down the avenues and into the subways with nary a care for the damage it caused.

Cooper's mind transposes these images onto his present view: a wall of water overtaking their platform, carrying him and Pando away down the tracks, never to be heard from again. The man in the trenchcoat with them is swept down the stairs and drowned, his white outerwear looking curiously like a lifebelt.

But wait — his focus returns — who is this man? This trenchcoated figure has appeared quietly on the platform, the only other person in the station. He must be waiting for a train, unless he isn't? Maybe this is a deadly Cokehead. Anxiety takes hold.

"I see him," Pando whispers, noting the man out of his peripheral vision. "Stay calm."

Lights can be seen reflecting off the silver rails at the end of the platform. Cooper pokes his head out and sees two bright bulbs growing larger in the tunnel. The familiar blast of air accompanies the train's arrival, with a loud rattling as it approaches the station.

He sees the oncoming lights as a means of escape — all it would take is one step into their path, and it'd all be over. He could jump right now; it's not like he has anything to lose. No more nightmares. No more hunger. No more worrying about the future. It would all come to an end.

He pulls himself back from the edge with seconds to spare. Tonight isn't the night for that. He has a job to do.

Pando joins Cooper alongside the train, careful to note the stranger with a series of subtle glances. The man doesn't seem to react to the train's arrival as it lumbers to a halt in the station. The doors open, and Pando's concerns subside as he watches the figure step aboard. He and Cooper remain on the platform as the doors close before them. The train convulses forward with a bump, then picks up speed as it glides down the platform and off to its next destination.

The danger past, Pando hops fearlessly onto the track below. "Let's go," he says.

Cooper, with one final scan of the premises, leaps down as well, careful to avoid the deadly third rail charged with a thousand volts. He flicks on the light of his phone and they begin their excursion down the tunnel, banking right for the City Hall Loop.

The hallucinations return to Cooper's mind; the waters are rising under his feet. His pace diminishes to a slog through the trash-strewn seawater, his ankles cold as he trudges along. Pando takes the form of another passenger to him, a fellow cabinmate doomed to the sea along with the rest of this shipwrecked world. He feels like he's on the *Titanic* once again, only this time it's New York that's taking on water.

The vision becomes too real when a roaring noise fills Cooper's senses. It isn't a hallucination this time; something is rattling the tracks. Cokeheads maybe, or maybe the tunnel really is flooding. Pando notices it too, and he turns around to check out their back flank.

"Oh *shit*," he yells, ogling a pair of lights making their way towards them.

Cooper whirls about to see the approaching danger. They haven't traveled far at all, but already it looks like the oncoming train is passing by the Brooklyn Bridge platform. The pair break into a sprint.

"We have to cross!" Pando shouts over the thunderous roar. They hop onto an adjacent track, narrowly clearing the electrified rail. Cooper extinguishes the light from his phone, and the two of them cover their ears as the mammoth train speeds by with earth-splitting power. Within seconds it rounds the loop, far ahead and out of sight.

Cooper turns the light back on. "Jesus," is all he manages to say.

"Let's hurry the fuck up," Pando urges, worried that more unexpected trains may come at any moment. They continue along the tracks, the phone light a meager protection from the danger of electrocution. One slip could send them to their death.

The underground loop is visible now, Cooper's light reflecting off the glasswork in the ceiling. Huge arches glimmer overhead, revealing the architectural splendor of the abandoned station. Pando pulls out the pocketknife as he spies a symbol taped onto the wall:

Cooper aims the beam onto the platform, and the two hoist themselves up, happy to be off the lethal tracks. In the white light of the phone, they see nothing but an abandoned station, cold and empty.

"This is it?" he asks.

Pando takes the phone from Cooper's hand, pointing it in all directions. "I don't see anything," he says. Then a flash of concern lights in his head: maybe this is a trick. Almost on cue, a clapping noise can be heard echoing through the cavern.

He angles the light in every direction. "Hello?" he calls.

A figure emerges from around the corner, palms slapping enthusiastically. "Well done, gentlemen," the familiar voice answers.

Pando trains the light directly onto the woman — her dirty blonde hair and bandana are a dead giveaway; it's Iri. He points the knife at her in self-defense.

"Are you gonna use that on me?" she asks.

"Iri?" Cooper asks, staring at her in disbelief.

"Congratulations," she says, stepping forward. "You passed."

Pando steps back fearfully. "What the fuck are you doing here?" he asks her. "You're working for Coke?"

She laughs. "Of course not," she says.

"Then what—" Cooper says, before being cut off when Iri raises a hand.

"There are no goods," she explains. "This was a recruitment exercise — and you both passed."

Pando aims the knife at her threateningly. "What do you mean, recruitment?" he asks. "I thought we were already recruited."

Iri shakes her head. "You think we let any average joe off the street become an agent for the Brooklyn Project?" she asks. "This is a war, Pando. We don't need volunteers. We need soldiers. And you two are brave enough to make the cut."

"So this was all just a test?" Pando asks, angry but relieved to not have to face down any Cokeheads.

"Correct," she says. "We had to test your loyalty, your strength, your commitment to the cause."

Cooper steps alongside Pando, getting a clearer look at the climate warrior before them. "So you're saying you have bigger plans for us?" he asks.

She nods with a smile. "You work for us, now," she says. "Welcome to the Clean Front."

Little Boxes

Herds of people travel in swarming currents through the channels of Penn Station. Cooper, heavy duffel in tow, is being tossed about by the busy humans, all bound for one destination or another in the dense holiday traffic.

Fuck these people, he thinks to himself as he bumps along, trying not to get trampled. It's so hard to love his fellow humans, he realizes, when all he wants them to do is get out of his fucking way.

Plowing through the torrent of bodies, he sees a large black man moving with him over the tiled hallway. The man's pace is brisk, but Cooper recognizes the opportunity. He moves in close behind to draft the walking wall, letting his broad body deflect any and all incoming humans that might retard his path. Hitching this ride, Cooper relaxes a bit, safely shielded as he makes his way to his bus to Florida.

Four months of city living has clearly taken a toll on Cooper's adolescent brain; all he wants is to get away from the bustle. And in less than a day, he'll be home for Christmas break, and again he'll eat like a normal person with three meals a day. Home. Not a moment too soon.

In his pissy mood, he leers at the humans around him disparagingly. Misanthropy colors his thoughts: look at all these idiots. They don't have a clue how the world works. Because if any of them did, they wouldn't buy into the illusions of this life. There's a man in a suit and tie, perhaps a

Wall Street banker ready to make a killing on the stock market. And there's a horde of Asian tourists, shopping bags in hand, marveling at the endless options of junk food. And here in line is a pair of fat human slobs, pushing along a stroller cradling another tiny fat human slob. They gawk at the puny creature, one more mouth they're only too happy to feed, no matter the cost. Isn't it all so wonderful.

Cooper takes another look at the plump baby in the teal stroller that takes up the space of two humans. Just another innocent born into a time of suffering, another victim of the systemic hypocrisy that allows society to adore the conception of life while simultaneously destroying it on a monumental scale. The misanthropy gives way to compassion and despair; his soul begins to weep for the small, helpless life. He pities any child under the age of five. They're gonna live in a world of tragedy. We all are.

He boards the bus and puts headphones in his ears, peering out the window as the engine roars alive. The fossilized towers of New York City quickly recede into the distance, and before long, all Cooper can see are trees — nature's skyscrapers. At last, he feels a degree of peace.

Old Nirvana posters cover the walls of Cooper's bedroom. They've been there since he was in high school, for he never felt the need to take them down. His room is quiet as he lays in bed, almost unsettlingly quiet, as he's grown accustomed to the noise pollution in New York. It's 7:22am in Kendall West; dawn light has touched the sky. Cooper rises early.

He crouches down to his duffel on the floor and fishes out a small plastic bag hidden underneath his pile of underwear. Inside is a blue lighter and a long, tightly-wrapped joint. Pando gave it to him so he can unwind while he's home; the fight begins again as soon as he returns.

The hardcover copy of *Futility* that his sister had given him for Christmas sits on his desk like a treasure. Meanwhile, his dad's gift, a shiny new smartwatch, lies toppled over next to it, still in its clear blister pack. Cooper isn't sure when he'll open it; the last thing he needs right now is another distraction.

His father and sister haven't yet awakened, giving him the perfect opportunity to take a relaxing walk. He tiptoes over the Christmas paper that litters the living room, sneaking out the back door into the cute suburban neighborhood he calls his home. From the safety of his back yard, he sets fire to the joint.

The house is an off-white cream, like every other house in his neighborhood. The green lawns interlace, the palms droop, and the early morning sunshine peaks over the red shingles that rise to the same height on every dwelling. In less than a century, the whole scene will be subsumed by saltwater, buried under the steaming, plastic-laden ocean of tomorrow.

Why can't I just live here forever, he thinks to himself. He didn't have a terrible childhood, by any means. Growing up without a mother obviously wasn't cool, but the act of living — finding food, having friends, excelling at school — wasn't the burden that it is now. When he was here, the possibilities of life were limitless. But now, as a poor starving adult in a city indifferent to his existence, life seems like nothing more than a day-to-day challenge. Sometimes reality is just too much to bear.

It doesn't help that his father hasn't yet realized that the transition to adulthood is nothing like what it was in his day. The millennials don't have as much to look forward to compared to the earlier cohorts; their forebears haven't left them as much. Yet these assholes complain when their kids live with them till they're 30 years old, moving into the basement because the world expects them to be something they're not.

They call us lazy, Cooper thinks, angry that his elders put all the weight of responsibility on him for cleaning up their mess. *But they're murdering us. We're not lazy. We're mortified.* He envisions scolding his father for pushing him out too soon. *We cling to you like frightened macaques because you left us unprepared to deal with this shit.*

With gold now filling the sky, the sight of his hometown appears a metropolis of spring. The Everglades are only a few klicks to the west, evidence to him that humans and nature can not only coexist, but flourish.

It really doesn't have to be one or the other. It can be both.

The little boxes line the empty streets, and Cooper takes in the view: the red white and blue soaring proudly atop a flagpole; a puppy barking somewhere in the background; gentle birdsong wafting through the air; a mail truck parked to one side of the road. It may seem bland, and perhaps too predictable, but it was never unsafe, and safety is the one thing Cooper wants more than anything right now.

The sprinklers spray happy fans of water over the grass, reminding the boy of this simpler life from a simpler time. The ire subsides; his muscles relax. It's a lovely day, and the suburbs are singing praises to the sun. This is the comfort of middle class life, the life his children will never get to know. Never know what it was like to eat fish for dinner. Never know what it was like to have running water on demand, to have a cup of coffee in the morning, to be able to fly anywhere in the world. They'll never know *this,* American suburbia. No human in history ever had it better than the people in these boxes.

But now everything's changed. In the fifty short years since his dad was born, the human population has doubled, and the wildlife population has halved. The suburbs are already in their permanent decline, and the race to the cities has begun. All that this life promised is turning out to be for naught. Now that he knows the true cost of this lifestyle, it doesn't seem so great after all. America is on the verge of change, not just climatically, but culturally as well. This way of life feels as ancient and ill-suited to the 21st century as hunting and gathering. Like all things in nature, only those who adapt will survive.

If you really loved us back, you'd understand, Cooper wails from within. *Maybe you'd even do something.*

A buzz from his pocket knocks him loose from his ire.

Dad: I thought you were going to put away the wrapping paper this morning. 8:02am

Cooper, realizing his transgression, wheels about and starts back for his home.

Cooper: Oh ur right, I'll take care of that right now 8:02am

He returns quickly, finding his father on the couch, sipping a cup of joe and staring at the mess on the carpet.

"Where did you go?" Jeremy asks.

"For a walk," answers his son, stooping down to grab the colorful scraps off the floor. Jeremy watches in silence as his son corrals the paper into his arms. He takes another sip of coffee.

"How do you spell *you're*?" he asks after a moment.

Cooper looks over to his father on the couch. "*You're*? As in *you are*?"

His dad nods.

"Umm. Y-O-U-apostrophe-R-E?"

Jeremy nods again. "Good. Just making sure."

Cooper rolls his eyes, returning to the Christmas wrapping. "You really think I'd go to Columbia and not know how to spell *you're*?"

"Well you didn't in your text," his father says. "I know you kids are too lazy to spell anything right these days. I worry for the world when you guys are in charge."

Cooper says nothing. He just continues to gather giftwrap off the floor.

"I think you should start your career after graduation," Jeremy says after a beat. "Grad school can wait."

The young man stands up suddenly, paper still in his hands. He turns to face his father.

"But if I go to grad school first," Cooper reasons, "then I'll have even more options when I finish."

His dad shakes his head. "You're so young, Coop. Don't commit to grad school now. There's so much time for you to figure out what to do with your life."

Rage pounces on Cooper like a leopard. He tries to resist, but the temptation is too great, the opportunity too perfect.

"*Is* there, dad? Is there *really*?" he mocks. "Have you even been paying attention? Because from everything I've been told, if I don't start doing something *now,* I won't have any future at all. So no, just because I'm young it doesn't mean I have time to kill. There's no time for that at all."

Jeremy scoffs. "You're too young to understand," he says.

"No," Cooper says, raising his voice. "You're too fucking *old* to understand."

Jeremy looks up at his son from the couch, surprised to see him lash out like this.

Cooper blinks, realizing he has his father's attention. "And you know what?" he shouts. "We don't need your fucking *you're.* Because we have *u-r,* and it does whatever we want it to. So keep your vowels and apostrophes and rules of grammar, because the rest of us are talking about how the fuck we're gonna clean up your mess." He throws the pile of garbage angrily on the floor.

"And from my perspective," the boy continues, "you look pretty damn lazy too. And selfish. And stupid. And I shouldn't be listening to you. I can do whatever the fuck I want with my life. Because I'm not your slave. I'm a fucking adult."

Flat Earth

Joseph sits comfortably at the end of a long, wedge-shaped table, with pairs of eyes facing him along its length. Behind him is a banner on the wall that reads "Profit is Progress," a window beneath it overlooking the Oklahoma oil fields under his dominion. The Coke Coalition is having a meeting.

"Arctic ice extent is at its lowest on record," Joseph informs the group as he flips to a map in his hands. "If the decline continues, we'll be able to deploy two additional processing ships to Chukchi as early as May," he relates with celebratory vim. The white men around the table nod approvingly, money on their minds.

Without warning, a trembling takes hold of the room. It starts of mildly, but then grows more powerful, rattling frames on the walls and shaking pens off the table. The Cokeheads look around calmly, waiting for the tremblor to pass. Within seconds, the quaking stops.

"Fracking earthquake?" asks a bald man with a hooked nose. He looks to Joseph worriedly.

"Could be," the old man replies.

"That's the third one this week."

"It's not important," Joseph shrills.

The bald man recedes in his chair, resuming a submissive posture. He continues to listen to his leader, with no further intention of challenging him.

"Lastly, we have an update on Project Chupacabra." Joseph turns the page in front of him, his pawns following suit.

"As you know, and investors are getting worried about the rising temperatures. "But we have to accept that the federal government may enact stricter emissions limits that would make our lives very difficult, and I'm not prepared to accept that, nor should any of you."

He considers his colleagues individually and deliberately.

The eyes look back at him. "So what do we do?" asks one.

"Each of you must implement a no-tolerance policy for using the phrases 'climate change' or 'global warming' in your businesses and administrations," Joseph prescribes. "We'll also continue to vote down any and all legislation that would regulate our industry. We can't bow to the demands of the greens."

One of the businessmen speaks out. "What about the climate treaty?" he asks. "They have enough votes; it would pass without any trouble."

"On that, we have a plan," Joseph assures them. "Dr. Valdez."

The aerosol specialist is seated to his left. They flip the next page of their packets in sync.

"A large payload of aerosols," the Mexican explains, adjusting his glasses. "It will be delivered into the stratosphere by specially-modified high altitude aircraft. Once the gases circulate around the planet, a global dimming effect will lower temperatures by a predicted 1C."

The men stare at their briefings with approbation.

"This is the solution we've been waiting for," says Joseph. "Our geoengineering investments are about to pay off."

"How soon?" asks a henchman.

"The prototype is under construction," he reports. "From there, we can begin full-scale deployment in a matter of weeks." He looks around the room with a deep shadow over his eyes. "I intend to use this before the climate negotiations come to a close. With the effect of this device, we can shut down any treaty in Congress with ease. No one will vote for emissions reductions if there's no need to reduce emissions."

A Cokehead smiles, appreciating the simplicity of the solution. "Brilliant work, Mr. Coke," says the old white man.

"Now I know I'm getting long in the tooth," Joseph begins. "I don't know how much longer I'll be around. But I know this — if the greens win the world, this country is doomed. And I'm not going to let them." He regards his cronies threateningly. "Any questions?"

Silence throughout the room.

"Thank you all for attending. Dismissed."

And with that simple word, they all rise to their feet at once, standing at attention for the departure of their master. Each of them has known Joseph for decades, but in this room, he is their supreme leader, and must be treated as such.

Joseph retires to his office down the hall, taking a seat at his rosewood desk. Next to his monitor is a framed photo of his family, children playing at his feet. One of them is a small Molly, beaming toward the shutter with a full set of baby teeth. Joseph's heart melts slightly every time he catches sight of her ridiculous pose, tongue out and eyes crossed. So innocent. So helpless. With any luck, she'll understand why he did what he's about to do. But she never will.

The bald man from the meeting appears in the hallway. He walks into the room unannounced, straight up to Joseph's desk, indignation on his face.

"Have you lost your mind?" the man implores.

The gentleman looks up at his compatriot with casual disinterest. "I beg your pardon," Joseph replies in his gravelly voice.

As the bald man is the Governor of Florida, Joseph has worked closely with him over the years to advance his political agenda. Though ideological allies and personal friends, the man is frequently one of Joseph's thorns, speaking out a bit more often than he would prefer. So when the Governor marches in and asks him, "You're actually going to do this?" Joseph's response is swift and resolute.

"Yes, I'm actually going to do this," he tells his minion. "And so are you."

A look of concern sweeps over the bald Cokehead's face. "You're willing to risk the future of the Coalition on this if it goes wrong?" he asks his boss.

Joseph glances down at the picture of his kids and grandkid. A twang of emotion flits through his heart — his Coalition is the least of his concerns. "I'm willing to risk everything on this," he affirms.

"Well I'm not," the Governor says defiantly.

Joseph looks at him with ire. "We've been friends for years, Rick. How many bad calls have I made?" he asks rhetorically. "Why do you doubt me now?"

Rick's face turns ugly. "Because I can't do my job as Governor if my people aren't allowed to talk about how to deal with Miami being underwater."

Joseph waves a hand dismissively. "Those models are inaccurate, and you know it," he says, denying the evidence.

Rick shakes his head in agitation. "I think you've gone too far with this," he opines. "The dignity of our party is in jeopardy."

His leader squints at him quizzically. "Are you up for reelection next year?" he asks; the Governor nods modestly.

Joseph tears open a drawer and pulls out some paperwork. "We'll see about that," he says dryly. With pen in hand, the powerful elder resumes his work, completely disregarding the man before him.

The Governor, recognizing the cue, departs from the room with a scowl. This small act of defiance could cost him dearly, and at the end of the day, winning elections is all that truly matters anymore.

Joseph looks up as he leaves, his eyes tracking with uncertainty. This is no time for insubordination. He has never been one to second guess his decisions, but with Rick challenging his judgment in such a manner — in front of the others, at that — the seed of self-doubt begins to germinate in Joseph's mind. Maybe he is wrong after all.

He rises from his chair and peers out his office window to the oil and gas fields beyond, their huge black shapes swinging up and down like scythes. With each cycle, they pump up another drop of sweet crude, another hundred calories of energy to power this sinking civilization. He notes the exhaust flowing up and away, the gas being flared off into the sky. Away it goes, to live in the atmosphere for centuries.

He turns around again to look at his granddaughter in the picture. Maybe it isn't worth it.

His concerns evaporate as he spies a familiar aerosol specialist approaching from down the hall. With a nod, Joseph signals the man inside.

The stubby Mexican doctor enters like a tornado, his feet thunderclapping against the white floor tiles with every step.

"Mr. Coke," he says.

Joseph leans back in his chair. "Come in, Jorge." He watches as his trusted colleague takes a seat. "Have you finished the program?"

Jorge pulls out a thumb drive from his breast pocket. The virus encoded on it will be able to track even the tiniest details of whoever is unfortunate to pick it up. It isn't hard to hack a phone.

"All you need now is a target," Jorge says.

"The counterstrike is in the works," Joseph replies. "The greens won't be a problem much longer."

Jorge smiles confidently, certain that his boss will deal with this nuisance. His attention shifts, presently gazing out the window at the enormous machines working busily against the red horizon. Billions of calories, all bubbling to the surface at once. The untold energy of a million sunlit days, stored like a bank account for the day when a species like us could make use of it.

"Someday they'll remember us as gods," Jorge muses. "Masters of the material world."

Joseph chuckles. "Or damn fools," he utters grimly. "But not if I have a say in it."

. . .

The Times Square subway station bustles frenetically as Cooper stands near the base of a ramped pathway. His territory has been staked along this walkway with two brochure stands on either end. The stretch of display between them includes picket signs with BKP slogans and large, diagrammatic posters along the walls. Rivers of people blitz by in an unbroken flow. Nearly everyone ignores him.

"Climate change is already here," he booms, throwing his full volume into the torrent. The people walk on by. He begins to wonder if this is all just a waste of time. Tick, tick.

The air in the underground station is undeniably arctic. While Cooper was south visiting home, winter arrived in force for the Northeast. Clouds of condensation emanate from every mouth and nostril like car exhaust. A weeping violin can be heard in the background, a heartbreaking dirge in this center of human machinery. The bodies rushing by are little more than drones, walking corpses funneled into currents like carefully programmed moving parts. Each face is locked in an unbroken stare, either directly ahead or at the slates in their hands.

One intrepid man is able to break the spell, disengaging from the forward momentum to consider the signs he otherwise would have passed.

"Hey buddy," he says, walking up to Cooper. "What's all this about?" The man is tall, wearing thick glasses, and has a prominent birthmark on his chin. It's the first person to talk to him all afternoon.

Cooper shines a welcoming smile. "I'm with the Brooklyn Project. Have you heard of us?"

The man's face morphs into a puzzling frown. "No, what's that?"

"We're an organization dedicated to fighting climate change through civic activism," the young man explains. "You know about climate change, right?"

The guy cocks his head. "Yeah, like global warming, something like that? I've heard of it."

Cooper offers him a pamphlet. "You need to read this," he says. "I think you should be prepared for it."

The man furrows his brow. After some trepidation, he accepts the brochure. "Huh. 'Climate Change Is Simple.' I'll take a look at it."

The boy is relieved to have some measure of success, no matter how small. "We have meetings every other week," he tells the stranger. "You should come to one. It's honestly a life-changing experience."

The man peruses the graphs in the pamphlet, especially the ominous Keeling Curve. "This looks pretty serious," he says.

"It is," agrees the youth. "But the good news is that we can still stop it. The more people like you join us, the better our odds of survival. You could be one of the people who save the world."

The man looks up at him. "Is that what you're doing by standing here?" he asks him.

"I'm trying to," Cooper says patriotically. "If I can help even one person understand climate change today, then I've done my job."

The stranger nods approvingly. "Well I'll keep this in mind. And your name, young man?" he asks, extending a hand.

Cooper accepts proudly. "I'm Cooper, Cooper Braydon."

"Nice to meet you, Cooper," says the stranger. "Keep up the good work." The man walks off, brochure tucked in his hand.

Cooper resumes his preaching, speaking of the horrors of imminent sea level rise, but the people keep walking by, oblivious to his presence. He turns to get a final glimpse of the potential recruit he just met. The pamphlet is no longer in the man's hand; it's lying on the floor in his wake, promptly trampled by innumerous pairs of feet. What a waste of paper.

On the streets above, a frigid winter wind socks Shree in the face as she shivers her way down the snowy sidewalk. The day she's been dreading has finally arrived. It's the first bubbling party of the year — an institutionalized gathering of teachers, sequestered in an apartment, to review their students' exam sheets and replace wrong answers with right ones. An afternoon of erasing and bubbling with #2 pencils, ritualized cheating over booze and finger foods. A celebration of standardized testing.

If she had known this was a common practice at her school, she never would have signed up for this job. But with government funding critical for their survival, the faculty can't take any chances. Their students must pass.

She crosses a fold of paper on the street, deposited there by the wind from some other place. Carried along in the glacial air, it lands only a few steps from a recycling bin. Her conscience, weighed down by the cheating she's about to undertake, sees this as a chance to add good luck to her karmic savings account.

Her body hinges downward as she grabs the paper, which she brings over to the green can on the side of the road. She releases it directly above the slot, leaving gravity to do the rest. A sudden gust blows it off to the side, missing the target entirely.

Frustrated, she bends down for it again. The wind persists, however, sweeping the white scrap halfway across the street. It flies aloft, further and further out of reach. No use now; just another urban tumbleweed flitting down the pavement.

After ten minutes of tundra, Shree reaches the destination apartment.

"You're late, Sharifa," says Principal Sanchez.

"I'm sorry, I can't control the trains." Shree walks inside and adds her heavy jacket to a pile on the coatrack.

"Now that you're here, I just want to reiterate the rules," Sanchez begins. "What happens here stays here. By the time this is over, no one will remember it ever having happened."

Shree looks at her spitefully.

"This is about your future as an educator, and our future as a school. These kids *will* pass their exams." The principal peers deep into the young teacher's ebon eyes. "Understood?"

"Yes," she responds meekly.

"Good. Go grab a stack of answers. There's sangria on the table."

Shree enters the modest living room, lit in warm colors and peopled by teachers. She finds it hard to believe how complicit they all are with flagrantly violating standards of educational excellence. People will do anything for money.

Bundles of tests are wrapped in rubber bands on a coffee table, boxes of pencils and pink erasers scattered everywhere. The crew is digging in dutifully, calculating how many answers need to change for each student to pass.

"Whose class is Diego in?" asks one, holding up an answer sheet.

"You can use my ID," replies another. They collaborate on how, specifically, his test results should be properly altered.

Shree looks around sullenly at the scandalous behavior, not wishing to involve herself. Sensing that the principal is watching her, she absently grabs a stack and begins looking through the answers bubbled onto the page. Most of them are wrong.

"Jesus," she whispers. She'll be here all day.

She looks at all the paper in the room, reams and reams of it. How many trees were slaughtered so her students could take these stupid tests, only for the adults to change their answers anyway?

Curious, she opens the calculator app on her phone to do some quick math. 20 metric tons of carbon emissions: the annual contribution of the average American. Multiply that by 23 years, give or take. 460 metric tons. That's Shree's contribution to date, her black mark on the world that she cannot erase. Divide it by 0.0015, the amount in metric tons a tree can inhale in a year. 306,666.666 trees. That's how many trees she's worth just in carbon alone.

And how many creatures lived in those trees? Figure half a dozen small animals per tree as the most conservative number, to say nothing of the trillions of microbes crawling on it. Then there are the trillions of microbes in the guts of those animals, not to mention the uncounted generations of every single organism with a chance to reproduce. She keeps adding layers to the calculation, the numbers proliferating wildly. The final total reads "Error" on her phone, too large for its memory. In this light, her one life looks very costly indeed. Maybe too costly; maybe not even worth it. She would rather the trees.

How much does a human life have to be worth before it's more than the world can afford? 500,000 trees? A million? This is a number she had better figure out soon.

. . .

Blurred fingers type out a brief message.

Cooper: How's the drone division? 11:12am
Pando: crazy man. its like im learning a real life video game 3:39pm
Cooper: You want to smoke later? 3:44pm
Pando: i smoked the last of my stuff last night, unless u wanna pick up an eighth from that place in brownsville 3:46pm

Brownsville. The hood. Not exactly Cooper's idea of fun. He decides to go anyway.

Cooper: ...fine 😶🔫 3:46pm

He descends the steps into the subway and loads money onto his metro card. A woman who looks like a pregnant stripper is flipping bills to the clerk at the station booth. She joins Cooper on the station platform just as a train pulls in; they enter the car together.

Cooper takes the nearest open seat, kicking two discarded chicken leg bones underneath a bench. What kind of person would just throw their shit on the floor like that? And who would eat chicken on a subway anyway?

The sound of a wailing baby penetrates Cooper's ears like a spike. There must be some hard-wired biological instinct that makes the sound of a child's distress so painful to the human psyche. Cry away, poor baby. Your suffering is just getting started.

The people on the train have their heads bowed at various angles, their eyes fixed on the screens in their hands. How he'd love to replace those phones with graphs, maps, and figures, and force the light of awareness into their minds. But in this life of instant gratification, if it takes longer than a 30-second internet search to understand, it's too much effort.

"Congratulations!" exudes a woman upon learning that her friend beside her is pregnant. She herself has a covered stroller in front of her, probably the source of the howling that grates on Cooper's brain.

Congratulations? he thinks. *Really? You're congratulating her?*

This is natalism at its finest. Because more humans is exactly what the world needs right now. You're pregnant? Congratulations! The miracle of life! Another homo sapien in the world, another average person of average ability that gets to take a way-above-average share of the world's calories, carbon, time. Another entitled American, eager to hijack the future. Another spectator to the mass famine, the refugees from California streaming from their homes to places not ravaged by drought. Another body to be enslaved to a phone. Yes, what a miracle, what a joy.

And the best part of all? They don't even realize that the world they'll leave their child will be a literal hellscape, a wasteland, a pile of ashes. And while no ethical person in their right mind would subject another human to such suffering, it's exactly the choice these parents have made, adding one — or maybe two or three by the time they're done — to the number of people who will eventually have to endure the most terrifying chapter of civilization. This child will never see a world of Holocene temperatures. This child will never know the abundance of the prosperous days of the age of carbon. Refrigeration, vaccinations, electrification — these things don't exist without civilization. And in a 4C world, there is no civilization. None at all.

But of course these parents don't see any of this. They don't see that this child is merely a tool for their own personal fulfillment, another selfish indulgence for the American way of life. After all, what couple is complete without parenthood? That's like a pocket without a smartphone, a river without a dam.

Will they ask us what we were thinking when we endowed them with life? Will they wonder if we even knew that by the time they reached maturity it would be too late, that the climate feedbacks would be in full swing, that the Rubicon had been crossed? Will they hate us? Will they wish we hadn't even bothered?

Oh happy times, another human birth to celebrate! Happy for everyone but the child, crying and screaming as it comes out of the womb, as if it somehow knows that even nonexistence might be preferable to the hand it's just been dealt. But no one dares reprimand a couple for choosing, in essence, to torture their children. All we talk about are happy times. So much to look forward to. Just not for the child.

The train feels like it's going to tear itself apart, rattling furiously down the tracks, its occupants swinging to and fro.

The sad thing is that none of our descendants will have any clue why it happened, for history ends with civilization. The humans that survive will eventually forget the sparkling world that was, subsumed by the desolation that is. And then Man will evolve once again. Maybe this time the smart ones will survive.

A woman appears from another train car, cloaked in a linty blanket that looks like the trunk lining of a car. A ripe, hazardous odor emanates from her body like gamma radiation.

"Somebody please help me," she moans as she moves through the morass of bodies. The deep lines on her face reveal a life clearly overwhelmed by the trials of existence. "No, don't turn away, please. Please, I need help. Will somebody please help me."

People stand awkwardly, waiting for her to pass by so they can get on with the commute they've made a thousand times. The middle-aged white woman in front of Cooper makes a sour face at an older, wealthier white woman. Neither of them even glance in the direction of the broken spirit slouching by, this struggling lump of humanity that for a brief moment brushes their backs. A human touch; flesh and warm cloth. Not a text message flashing on a phone, but body-to-body contact.

"Please help," the woman calls again. What clearer a plea could be raised to a crowd? This isn't a complex message — it's not like explaining arctic amplification or agricultural methane emissions — it's just a clear SOS. It's the easiest thing to understand: I need help. Just lend a fucking hand, *please*; this life is too much.

Cooper doesn't look up. He feels her pain though — he gets it. Sometimes this life *is* too much. Sometimes death is the easiest answer. But it's rarely the right answer.

"I need help," the Earth cries again. But we the people have our headphones in our ears, our eyes buried in our phones, never making eye contact with the stricken, suffering life in dire need before us.

All she wants is a dollar. A smile. A warm bed. She wants a morsel of food, and perhaps a thimble of water. She has basic needs that aren't being met, likely through no fault of her own, but the fault of a people that choose to ignore the obvious, people that discriminate and draw battle lines, when such distinctions are so unhelpful that it only serves to diminish the humanity of the whole.

None of these people, Cooper included, can be counted amongst the ranks of humanity in this moment, instead surrendering themselves to a mentality of nonaction, because even the slightest inconvenience seems too much to sacrifice. The woman will die alone, giving up on the world as the world has given up on her.

At last Cooper departs the train, the chilly wind slashing deep into his pores like the teeth of a great white. He misses Florida.

A plastic cup scrunches under his boot as he traverses the frozen landscape. He sees four smokestacks belching wispy streaks into the cobalt sky, billowing away merrily, unaware that every promise built into its bricks will soon be broken. The pillars remind him of a certain liner.

He sees the stoop in the distance, shady men guarding it like eagles.

Realizing he might look too pathetic to stroll up to them asking for drugs, Cooper ruffles his hair, lowers his pants, and flips his hat backwards.

"Sup guys," he says casually. The men are much larger than him, muscles bulging through their sweatshirts. They could hurl him across the street like a javelin if he crosses them the wrong way.

"Whatcha need, boss," says a man with a spidery tattoo on his face.

"An eighth," says Cooper in a low voice.

The man nods. "Wait here." He disappears into the building, returning moments later with a small plastic bag. He hands it to Cooper discreetly.

The boy quickly pockets the bud. In turn, he folds him a fifty, which the man hides in his sleeve.

"Thanks."

As quickly as he came, Cooper leaves. So easy, and very professional. 10/10, will do business again.

On approach to the subway, he spies a handful of cops hanging out by the turnstiles, as if they're waiting to apprehend him. He freezes in his tracks at the sight of them, four large men, broad like tombstones.

He climbs back to street level with paranoia. He's never carried drugs on his person before, let alone past a small army of police. He decides he better not take any chances, lest he attract any undue attention. He doesn't want to be stopped by cops, for any reason. They shoot people.

He tucks in his shirt, pulls his socks high as they'll go, and orients his hat bill-forward. There; much better. Now he looks as white as a Mormon. Though he stands out even more against the citizens of this hood, at least he knows he won't be profiled.

By the time he reaches his apartment, a minor party is in full swing. Marquis is over, playing a video game with Shree. Their phones act as controllers, the game being viewed from a tablet balanced on the living room table.

"Oh, damn, get touched!" Marquis shouts at the screen. "That was so dirty! What a cake transition."

Mel and Pando turn to Cooper as he shuts the door.

"Thank God," says Pando, excited not just for his friend, but for weed. He and Cooper greet each other with a fist pound.

"I thought you didn't believe in God," Cooper goads.

"Fine," he says, taking the bag. "Thank *you.*" Pando loads a bowl and draws a long, deliberate hit.

"OMG, respect," blares Marquis, relinquishing his controller. "Good match."

Shree grins smugly. "Let's celebrate."

The air inside the apartment is chilly, for want of better heating. The living room is decked in white Christmas lights, a galactic haze covering the walls. Math rock plays from a phone, drifting throughout the space. Cooper notices that Shree is dressed in black from head to toe, defying the otherwise buoyant atmosphere.

"What's up with you?" he asks her.

She exhales a sooty mist. "I'm in mourning," she says, passing off the pipe.

"For what?"

"Everything."

"Oh."

Mel takes the bubbler from her. "She's just sad because the world is ending," she says, raising the pipe in the air. "A toast: to the future." She inhales deeply.

"Fuck that," says Pando. He takes the piece as soon as she's done. "To us: the final generation of the golden age. A time when men were gods, held the world in their hands, and crushed it flat. Generation Weed: committing the world to a slow burn." He takes a hit.

A bug whizzes past Mel's face, tiny and brown, with huge red-orange eyes. She slaps it between her palms, watching it struggle as it dies, a puny irrelevance in the big human world.

"There are flies in here," she complains. "We need to fucking clean this place." She rises for the trash can. Upon opening the lid, a swarm of insects darts out. She shoos them away with her hand.

"Why are you changing the trash?" asks Pando.

"Because it smells."

Pando whiffs the air loudly, nothing but the scent of ganja entering his nose. "I don't smell anything. Do you smell anything, Cooper?"

The boy averts his eyes, hoping to stay out of this one. These two do not know how to play nice.

"Whatever, I'm changing it anyway," Mel says.

"It's not even half full yet!" He's appalled by her flagrant wastefulness, her incessant whining. He's watched Mel leave messes more times than he can count. Why should he have to clean up after her? The day she picks up a broom will be the day he makes an effort. And he knows that day will never come.

Mel peers into the wastebin, spying a half-eaten burrito in the bag. Probably the source of the flies.

"Remind me to finish my food before I throw it out next time," she says, tying the plastic into a knot.

The thought of uneaten food insults Cooper. His stomach is an empty hole.

"I'm so hungry, dude," he says. His belly chimes in, corroborating his claim.

"So eat something," says Pando.

"I can't, I've already eaten today."

Pando grabs a bowl of grapes from the kitchen and offers it to his friend. Cooper stares at the green globes fluorescing in the white light. Their tempting plumpness activates some instinct deep in his primate brain. With his thumb and index finger, he reaches for a grape, squeezing it slightly to gauge its tautness. Satisfied, he plucks it off the vine and puts it in his mouth. He chews, the cells dumping fructose and other chemicals straight onto the most salient parts of his tongue. Their effect is intoxicating; he rabidly picks more fruit off the vine, stuffing each seed into his mouth, delighting in nature's candy like a simian in a tree.

"That reminds me," Pando says, watching Cooper eat prodigiously. "I have more food for y'all."

He heads for the fridge, pulling out a large white bag and unloading its contents into the microwave.

"Oh, wait," says Marquis. "I just remembered I brought weed too."

The microwave chimes, and Pando returns with a hot sack of food.

"Here's five thousand calories," he says, putting it on the table with a stack of plates. "Dig in."

Cooper does the math in his head. 5,000 x 10 = 50,000 calories. That's how much fossil fuel energy this bounty burned.

"We also have chips," says Mel, rummaging through the pantry. "And we have beer."

"And we have vodka, and gin," says Marquis, pulling out all the things he forgot he had in his backpack. "And weed. We have a fuckload of weed."

Pando claps approvingly. "Oh shit, well let's just fucking kill ourselves!" he delights.

Cooper stuffs his face with a plate of meat. He gets so focused on the experience of eating that he loses awareness of his body. His entire world shrinks to just his mouth and the food.

A buzz from his pocket distracts him. He reaches for his phone, a text message from some unknown number:

Unknown: hey 6:21pm

He looks at it probingly. *Hey*? That's it? He starts to tap.

Cooper: Hey who's this? 6:21pm

The world around him falls away, his focus now on his phone. Mel and Marquis start playing video games; Shree and Pando smoke more. But who is this mystery texter?

"You guys recognize this number?" he asks his friends. No one seems to know who it might be; they all go back to what they were doing.

Another text message. This time it's a picture, a selfie of a stunning girl, though one he doesn't recognize. He taps on the photo zealously, embiggening it to full screen, inadvertently infecting his phone with a malicious package of spyware. He texts her back.

Cooper: Do I know you? 6:24pm

It doesn't matter that he doesn't know her. He wants to. Immediately.

Marquis, deeply concentrating on the video game, begins beatboxing to himself to rhythmically focus his mind.

"That's a sick beat," Pando muses. His mind turning over in his head, he starts feeling the rhythm with a head bob.

"They say we know jack, but I say they whack," he spits. "Our lives are worth more than just another world war. They kill us in the streets, beat us head to feet, but they just can't see that we revolutionary."

Marquis beatboxes more intensely now, feeding off Pando's exuberance. He's feeling it.

Cooper watches as Pando raps about the injustice of the world, belting out a beautiful, lyrical poem about how we want our future back. And it's in this moment that Cooper sees, for the first time, definitive proof that Pando is, undoubtedly, one of us; that he is without question among the ranks of humanity and not some misprogrammed robot. He's far from some mere fatalist like Mel would have him believe. He's just a tortured soul, longing for expression. From his rhymes alone, it's clear that Pando isn't merely an angry young man, but rather something far more profound: a savior. Pando could, in fact, be the answer. His lyrical genius. Commercial gold, and bitter as hell. This generation's Bob Dylan. Pando Lamar.

"Damn, man, that was good," acclaims Marquis. Shree smiles happily, rubbing Pando's black hair.

"How do you know all that shit?" asks Mel.

He smiles pompously. "Because I read the news. You should try it sometime."

She laughs sharply. "Maybe I will."

Marquis, in a moment of clarity, exclaims a whooping "Oooh!" as if he had just unlocked the secret to cold fusion.

"What?" demands Cooper, hoping perhaps he knows the identity behind the number on his screen.

The young man turns slowly to Cooper. "I'm high," he confirms.

Cooper falls back onto the mandarin cushion, unamused. He unlocks his phone again, checking to see if the girl has replied. His fingerprint scanner takes a moment longer to recognize his signature than normal, but he doesn't think anything of it. No new messages.

Whatever. It's not important, he decides. What's important is he's having fun with his friends. They're all so high, enjoying the high times of humanity.

The Logical Song

Careful not to fall, Shree makes her way down from the loft at a deliberate pace; her low-heel shoes aren't ideal for traversing rope ladders. But there are eggs in the fridge calling her name, and she responds obligingly, school materials in hand. As she descends, she notes the man typing frenetically on the couch.

"You're up early," she says to him.

"Just getting some work done before I go to BKP later," Pando replies.

Shree frowns as she continues into the kitchen. "You sure have been spending a lot of time there lately. What are you guys up to?"

Pando smiles at her. "Umm saving the world, duh," he says, without elaborating. He's been putting in 15-hour days to prepare for an upcoming recon mission at the Cokeheads' geobomb compound. For months he's been training to pilot a drone, an assignment he optioned for immediately, it being the most intriguing role he could sign up for. In a couple weeks it will all pay off, and he'll fly right over that building in Mexico, getting one final closeup before the big strike. His mission, so vital for the next phase of operations, is all that remains between now and the revolution.

Shree scoffs at his answer. "Right. Cuz we Americans are all about saving the world."

"Obviously."

A door flings open against the wooden loft railing; Mel emerges from the crevasse in a crawl. Nothing but a beige towel covers her pearly skin.

"Morning," she grumbles to her roommates below. With the groggy expression of a sleepwalking zombie, she descends the ladder and shuffles into the bathroom. The sound of water pouring through the showerhead can be heard through the door.

Shree combs the fridge for the eggs. Three treasure chests of golden runny yolk should be there on the top shelf, but there's no sign of them. After a minute of searching, she sighs bitterly and looks for alternatives; clearly there won't be eggs for breakfast today.

The bathroom door opens again, the showerhead still running. "There's never any fucking hot water," Mel complains. "Like why? Why can't I just have five minutes of hot water?" She looks accusingly at her roommates, as if they siphoned off the heat before she had a chance to enjoy it.

"First world problems," Pando utters. He continues to write on his computer.

Shree pokes her head up from the fridge. "Have you seen my eggs?"

"I had some eggs last night," Mel answers. "But those were my eggs."

Shree's brow scrunches unhappily. "Were they in the yellow carton?" she asks.

"I think so," Mel says dully.

"Then those were my eggs," Shree frowns. Disappointed, she reaches for the half gallon of milk on the door shelf. She shakes it pathetically; there's barely enough left for a bowl of cereal. "Cap'n Crunch it is," she supposes.

Steam pours forth from the open bathroom; the water is piping hot. Mel, sufficiently appeased, returns to the shower.

Shree takes a seat on the couch next to Pando. She chews her sugar cereal loudly as she browses the headlines from her phone's news app.

"So the funding bill passed," Shree says unhappily, the legislation in question curtailing vital resources from education to other programs.

Pando nods. "Did Coke vote for it?" he asks her.

Shree scrolls through the list of votes; Joseph Coke is one of the first names. "Of course he did," she says. She shakes her head angrily; enmity grows in her heart. "I still can't get over how a guy that doesn't believe in science is the leader of the science committee."

Pando laughs. "Welcome to America: land of the free, home of the dumb. Our leaders are asleep at the wheel."

She knows he's right, but the fatalism isn't helping. Despite her outward positivity, her life has always been marked by a certain emotional fragility.

"I think I'm addicted to misery," she wonders aloud. "Whenever I read the news, it makes me so frustrated, to see all the bullshit happening that I can't do anything about." She looks to Pando, who seems to agree. "I feel like shit about it, but it feels good to feel bad, because I *should* feel bad about where things are going, right?"

"You mean to hell?" he asks. "Yeah, I'd say so."

She tries to maintain a spark of optimism, the kind that kept her alive through all the trials she's faced till now. "Well, at least there's the climate treaty in May."

Pando scoffs in rejection of this meager shred of hope. "So what?" he jeers. "It means nothing."

She looks at him disgustingly. "Why? Why does everything mean nothing to you?" she storms. "Why can't you just feel a little bit of hope?"

His face turns severe, his eyes shooting silver beams right through her. "Because what has hope done?" he snarls. "We've been *hoping* for years. When it comes to climate change, hope is an extension of denial."

He opens a new tab on his browser and pulls up a webpage with the official goals of the forthcoming negotiations. "Look at what they're trying to do," he says, showing her the details. "It won't be enough to stave off the worst. We'd still be on track for 4C. So no, this treaty doesn't give me hope. It's a colossal waste of the time we don't have."

Shree sighs at the inevitability of his outlook. "I for one am happy to see people negotiating," she says.

Pando returns to his screen. "They've been *negotiating* our whole lives, Shree. I think it's time we actually *did* something."

The bathroom door scrapes open. A cream white figure materializes from the milky cloud inside and hikes back up the loft. Shree rises as well, taking her empty bowl to the kitchen and grabbing a Lunchables from the fridge.

"I'm off for work," she announces.

Pando puts the laptop on the table and steps up to hug her goodbye. "Have a good day," he tells her with a grin.

"I'll try," she says, weakly returning the smile. The door closes behind her.

Mel dresses hastily, and in minutes back downstairs, leaving without acknowledging Pando on her way out. Half asleep as she boards a train, she recalls Pando's dare from last night — to read the news. Not one to pass up a challenge, she pulls out her phone and taps the news app. She flips through the headlines in the fluorescent light of the train.

"Gunman Opens Fire in Louisiana Restaurant, 11 Dead." She swipes her finger across the screen, revealing the next headline. "Latest Jobs Report Weaker than Expected." Again her fingers slide over the glass. "US Begins Arctic Oil Drilling." Swipe. "Terrorist Plot Foiled in London." Swipe. "Worker Killed by Robot at Teddy Bear Factory." Swipe. "NASA Declares January Hottest on Record." Swipe. "Suicide Bombing in Afghanistan Kills 43." Swipe.

"Man this shit is depressing," she grumbles to herself. She considers the next headline. "Cutest Puppy on Earth Will Make Your Day." She taps the screen. Finally something worth reading.

"I pledge allegiance, to the flag, of the United States of America," the students chant religiously. Shree watches as every one of them praise the red white and blue, standing at attention in the stuffy, overbearing gym. Allegiance to a nation unwilling to invest in the future, preferring instead to stick its children in this sweatshop than teach them anything of value. A country whose lust for oil and power transformed the world into the greenhouse it is today, a footprint the size of the planet itself. And in the twenty seconds expended in saying these empty words, a thousand tons of carbon have entered the atmosphere just from America alone. "With liberty and justice for all."

She closes her eyes tightly; she has to put the negativity out of her mind, and denial is such an easy means of escape. Maybe America isn't a plutocracy in the grip of carbon polluters. Maybe she isn't complicit with institutionalized cheating. Maybe, just maybe, life isn't as bad as it seems.

The stack of altered tests answers on her left desk suggests otherwise. She grabs the pile and begins distributing them among the class, telling herself that cheating is a necessary evil; better to have the money to not teach them tomorrow than to have no money at all. She returns to her teacher chair with bleary eyes.

A student rises from her seat in surprise. It's Trina, the sweet little girl that can't seem to figure out up from down, but tries so damn hard to understand. She starts for Shree's desks. Perhaps she's onto something suspicious.

"Miss Dunlap, look!" the girl says, brandishing her test in the air. "I passed!" she exclaims, flashing her crooked teeth. "I can't believe it!"

This should be a big moment for the budding instructor, the first time one of her students feels like they've excelled at something. For a girl that fails almost every test, this must be a moment of pure elation. And as her teacher, Shree can take credit for her student's stunning success. It crushes her to the core.

"I can't wait to show my mommy!" Trina proclaims, jumping up once and then rushing back to her seat. Shree does everything she can to hold back the tears.

. . .

Work has been long for Cooper. His Straunt uniform smells of coffee and grilled cheese; closing shifts are never fun. And now he has to travel home down the desolate streets of 3am Brooklyn, knapsack bouncing against his back, hunger pangs biting his stomach the entire way.

The young man makes his way into the only lit convenience store along his route, an empty 24-hour chain with fully-stocked shelves of empty calories and cheap products. Acting on the whim of his physiological urge to eat, he grabs a colorful plastic bag of garbage from the snack aisle. It may just be a serving of chips, but it will quell his hunger till tomorrow, when he can grab a bagel while working the morning shift.

There are no cashiers on duty at this hour; the only checkouts available are automated. Cooper approaches one of these boxes, expecting to be recognized by the device's motion sensor. To his dismay, however, the machine doesn't react.

He steps forward, drawing his face closer to the screen. "Hello, computer," he says to it gently. Its screensaver runs idly, still unaware of his presence. Annoyed, Cooper pushes his finger into the glass; the register jolts itself alive.

"Welcome," the machine says politely without skipping a beat. "Please scan your items," it commands in its perfectly enunciated English, running through its limited set of preprogrammed phrases. Cooper pays for his chips and leaves, grabbing his unnecessarily long receipt from the mouth of the machine. Nothing but coupons on the back of this arm-length paper, discounts for products nobody needs, to be discarded into the landfills nobody sees, just more waste to bury this world that nobody cares about.

He resumes his purposeful walk home, tearing open the bag and munching heartily on the red chips inside. The crumbling sidewalk under his feet forms a now-familiar pattern of cracks, indicating that he isn't far from the apartment.

The loud crunching of the tortilla chips blunts Cooper's sense of sound, and he doesn't react with any sense of concern to the growling approach of a giant metal hound behind him. It isn't until the truck pulls beside him that he realizes what's happening, and by then, it's already too late to prevent. For from out of this vehicle hop two men in leather jackets, black like a seam of coal. Two flashes in the yellow streetlight indicate they're both packing heat.

Cooper's fear response activates with speed, prompting his body into a sprint, bag of chips still in hand. But the uneven sidewalk catches his shoe, sending him stumbling as the dark figures give chase. Within moments they are upon him, restraining him with their thick monkey hands as the boy struggles in resistance.

The taller of the two attackers socks him flat in the gut, subduing him immediately. The bag in his hands falls to the ground, scattering red triangles across the sidewalk. As the shorter man holds Cooper's arms behind his back, the other grabs the boy's hair and pulls him up, allowing the kid to look his enemies straight in the eye.

"I have a message for your leaders," the tall man says.

His anxious eyes struggle to notice the details: thick glasses, mole on chin. A surge of hate boils in his mind, the chemical brew of cortisol energizing his body's defenses with an upwelling of rage.

"If you do anything to get in our way," the man whispers, "carbon is going to be the last of your problems." His blue eyes pulse like a laser sight trained squarely on the frightened adolescent's head. With but a glance to his accomplice, Cooper is dragged into the middle of the empty street. The man in glasses returns to the truck, while the other throws the youth to ground, kicking him swiftly in the ribs with a brown pointed boot.

Cooper groans against the cold tarmac, his crumpled body limp on the ground. The man steps over him and returns to the rumbling truck, having taught the lesson he was sent to instruct. The vehicle revs up, making a sharp U-turn before driving off in the opposite direction. Cooper, still in the street, watches as the truck turns around again, its headlights now aimed at his body. With blinding speed, the beast accelerates straight for him. Helpless to move, he watches as its hungry lights grow larger in their approach. He accepts it without even a flinch.

With a second to spare, the truck veers to the left, avoiding him by less than a foot. A cloud of black smoke rolls over his body, covering him in a thick layer of soot. The demon speeds off into the night, its engine bellowing like a panther the entire way.

Echoes

An air of desperation hangs in the presentation room at BKP. Iri has pulled Nevaeh aside, speaking in hushed tones as she explains the situation. Nevaeh's face blanches noticeably as her she considers every word. Cooper watches from his seat on the bench, Pando by his side.

Nevaeh retreats to a back office with Cooper's cellphone as Iri returns to the table. She sits opposite the men, hands clasped.

"We're gonna figure it out, Coop," she promises. "We won't let them hurt you again."

Cooper's eyes implore her for more. "How the fuck did they find me?" he asks, hands cradling the bruises on his rib cage. "Are they gonna kill me?"

"They're gonna kill everyone," Pando scowls.

Iri lays her palms up on the table; Cooper, recognizing the gesture, puts his hands in them.

"Cooper," she says softly. "I know you're scared right now, but I promise you that everything will be alright."

"I don't think so," he says, shaking his head. Every day he learns new things to fear. He once thought growing up would help him make sense of the world so he wouldn't be afraid, but now nothing makes sense. He's more afraid than ever.

He withdraws his hands and covers his face, trying to collect himself under the intensity of his stress. Tears evade him, but fear and sadness consume his thoughts like a Javan Tiger. Wait, they're already extinct. Like a pack of wolves.

Pando stares at his friend with eyes of concern. He puts his hand on his shoulder, only for the adolescent to brush it off with a roll.

Iri walks over to a metal shelf and pulls off a shortwave. She flips it on and tunes it to a specific frequency. A freakish, clattering noise sounds from the speakers, chaotic but repetitive, like a demented piece of musique concrete. The broadcast rattles Cooper's frame like a blast of cold air.

"This is the Cokeheads' numbers station," she says. "This is how they're communicating with their agents, just like we communicate with ours." The screeching broadcast continues, nails on a chalkboard. It sounds like pure evil.

The sound grates on his ears like sandpaper. He shudders unhappily as he shakes his head in disgust.

"I don't think I can do this," the youth admits, his hands falling to his lap. "I'm having second thoughts about being with BKP."

Her jaw clenches. "Cooper, we need you," she pleads. "We're about to go fangs out. We need all hands on deck."

Cooper's trembling lip curls into a frown. Is he really willing to risk his life like this? He sighs heavily as he turns to Pando.

"We'll keep you in the loop," Iri promises.

As they walk back to the apartment, Pando looks at his roommate with eyes of concern. The boy really seems broken this time, the way his hands are jammed into his pockets as he gazes at the ground.

"Hey Coop," he begins. "I know you're feeling pretty low right now, but I think I have something that'll help."

Cooper grimaces severely. "What drug do you have for me this time?" he asks.

"You'll see," Pando answers with a smirk. Cooper shakes his head with amusement. He's already tried weed and finds it as great as alcohol, so what else lays waiting for his discovery? He may as well find out.

Pando has the tabs stashed away in a scrap of aluminum foil. He takes a seat on the futon and removes them from their wrapping as Cooper sits beside him. They've returned home at that magic moment in the day when sun trickles through the window, raining golden light into every corner of their box.

"You'll like this," the refugee promises his friend. "Whenever you need to step back and see the big picture, just take this." He holds it up for the boy to see; instead of being green, this drug is on four tiny squares of perforated blotter paper, each laced with the potent psychedelic known in some circles as Hoffman's potion. He breaks off a pair and hands it to Cooper, dropping them into his palm.

"Why aren't we doing this in, like, Central Park or something?" asks Cooper. "That would be like, the most incredible experience."

Pando shakes his head as he unfurls a bedsheet over the couch. "Nah, man. Set and setting. We wanna control the trip. Here we have music, food, all the comforts of home." The cloth settles gracefully over the sofa.

"You ready, man?" he asks the white kid. "This might be a doozy."

"I'll start with one," Cooper suggests. He looks around. Water bottle: check. Streaming music and speakers: yes. Window is locked, as is the door. They're ready.

"Alright, let's go," says Pando, a swashbuckler embarking for new shores, an explorer from the Age of Discovery traveling to the antipodes. He pops in the tabs and puts them under his tongue, absorbing them harmlessly through the membranes of his mouth.

Cooper follows suit, taking half the dose. Now it's time to wait.

"We have an hour to kill," Pando says. "Want to play video games?"

They don controllers and begin to race on a digital screen. Seven minutes pass; another species goes extinct. Then another, and then two more. And before Cooper is even aware of it, his perception is beginning to curl around him absurdly. There's acid in his blood, turning his mind silly. Now he can barely focus.

"I have to stop," Cooper says.

"Alright, no problem," Pando replies, pausing the game. "You feeling it?"

"A little bit, I think." He leans back comfortably, stretching his body across the length of the couch. His mind is traveling down new avenues.

Maybe climate change is a no-win scenario, he wonders. *Maybe we need to redefine the problem. Save the planet? No; save the humans.*

Captivated, Cooper launches to his feet. He approaches the slanting light of the window; outside he sees the people walking down the street below, tiny creatures on a giant breathing planet. Even in their myriad forms, the primates all look the same to him. The appreciable differences are superficial and irrelevant. Fundamentally, they're all human.

We're all the same story, he surmises, *just being told and retold; the same genetic book, reimagined with every new person, but variations of the same idea. The story of humanity. It's in all of us.*

Cooper's neurons fire rapidly; his pupils are at full dilation. The neural networks of his brain react to the hallucinogen, reorganizing themselves into new channels of thought. His body slumps into the couch, loosening into a puddle that soaks into the fabric. He feels himself melting, merging physically with the sofa into which he collapses. His arms meshing with the upholstery, growing into the couch like mycelium infusing fresh soil.

Cooper can feel his corporeal existence unraveling. His mind unlocks; the filters lift. Out comes a flurry of integrated thought, a holistic experience of perception the likes of which he's never known. His thoughts flit with frightful speed, disengaging from all preconceived notions and synthesizing a new paradigm out of their shattered remnants.

He's afraid to let go of his old ideas. His body is so important to him, his ego, his individuality. But he's being ripped from reality, and he's about to witness a new one from a new vantage. He'll see things in a whole new way, if only he'd just accept it.

The tornado in his mind blows furiously in its effort to scoop him up, but he obstinately resists. He clings to his blanket as the winds swirl through his childhood bedroom, the hail pounding violently through the broken window. His mother is calling for him. She's in the tornado this time; the tornado is her. And she's beckoning him to join her with the rest of the living and the dead.

He doesn't want to go, but the torrent is too strong. He's been here before; he's back in that night fifteen years ago, but he's here in the present too, torn between two moments in universal time. His mind spins violently, splintering apart in the unrelenting chaos.

The twister washes over him. He's cast aloft, the great force snatching him from the comfort of his tiny bed and tossing him high above the world, allowing him to see the it all for a few seconds before he succumbs to gravity. He closes his eyes, awaiting his imminent end.

But the howling subsides; the debris settles. And Cooper continues to float up, an angel rising into the breaking clouds. Realizing the danger has claimed him, he accepts his own death. He no longer fears it. And so he opens his eyes, conquering that fear.

A flood of light overcomes his vision, the purest white in all the universe. It blinds him, until slowly, gradually, the fog clears, and through it he sees . . .

COLORS, SHAPES, AND PATTERNS OF NO COMPARABLE BEAUTY. A COLLAGE OF IMAGES, THOUGHTS, AND FEELINGS GLUED TOGETHER IN NEOTERIC WAYS, RECONSTRUCTED INTO A COMPLETELY NEW WORK, A NOVEL CONCEPT. HE STEPS THROUGH THE DOOR, FLOATING IN A MOST PECULIAR WAY; THE STARS LOOK VERY DIFFERENT TODAY. A PANDEMONIUM OF NONSENSE DISCHARGES FROM HIS MIND THAT SEEMS TO MAKE SENSE OF EVERYTHING, OR MAYBE NOTHING AT ALL.

THEN HE WINKS OUT OF EXISTENCE.

. . .

. . .

. . .

SUDDEN REAPPEARANCE. THE HALLUCINATIONS OF TOMORROW FLING HIM TO A LAND STRANGER THAN KNOWN, THE MOUNTAINS OF A MOON FAR, FAR AWAY, WHERE CASTLES IN THE SKY FLOAT IN MYSTIC CLOUDS OF PURPLE HAZE. THESE SHALLOW IMPRESSIONS MATERIALIZE BEFORE HIM INTO A UTOPIA OF SOFT SKIES AND PURE NATURE.

THIS BLUE MARBLE IN THE DARKNESS IS HIS HOME. HE WATCHES THE RENAISSANCE FAIR OF HUMANITY PLAY OUT ITS DRAMA FROM ABOVE. PLANET EARTH IS BLUE, AND THERE'S NOTHING HE CAN DO. HE DISAPPEARS AGAIN.

. . .

. . .

BUT NO, HE MUST RETURN. THE WAR IS ABOUT TO BEGIN. HE NEEDS TO FIGHT.

TIME UNFURLS LIKE A SCROLL, EACH MOMENT IN HISTORY REVEALED BEFORE HIM IN UTTER CLARITY. HERE HE IS IN THE GAY 90S, WHEN TELEVISION CARTOONS DEFINED HIS LIFE. AND HERE HE IS A TEENAGER, A WHITE RABBIT ROVING LIKE A BABE IN THE WOODS. THERE'S A HAWK OVERHEAD, HANGING MOTIONLESS UPON THE AIR. HE REMAINS UNAWARE OF THE DANGER IN THE SKY.

A FLASH OF DELIRIUM SNAPS HIM BACK TO THE PRESENT MOMENT OF FOREVER, AND GOD COMMANDS HIM TO LET THERE BE MORE LIGHT.

HE READS THE NEWS TODAY, OH BOY. IT'S RATHER SAD. THE GLACIERS RETREAT, THE OCEANS SEAR. THE SIBERIAN SOIL BREAKS, SPILLING LOOSE ITS TOXIC METHANE. WE'VE GOT TO GET OUT OF THIS PLACE.

THE TREES BROWN, THE LAND DRIES. BUT PEOPLE ARE STRANGE, AND THEY TELL THEMSELVES OVER AND OVER AND OVER AGAIN THAT THEY DON'T BELIEVE THEY'RE ON THE EVE OF DESTRUCTION. GOD ONLY KNOWS WHAT IT WILL TAKE TO OPEN UP THEIR MINDS, BUT THE TIDE IS RIGHT TO TURN.

THEN SILENCE. DARKNESS.

. . .

. . .

. . .

. . .

. . .

. . .

THE MAELSTROM OF THOUGHT RESUMES ITS FURY. COOPER IS STRUNG THROUGH IT LIKE CHEESE THROUGH A GRATER, HIS BODY SEEMING TO DISTEND ENTIRELY, LAYER BY LAYER AND SYSTEM BY SYSTEM: SKIN, MUSCLE, LUNG, HEART, GUT, BLOOD VESSELS, BRAIN, CELLS, ATOMS, QUARKS, BOSONS. HE JOINS A UNIVERSAL ONENESS IN THE COSMOS, AN EVERYTHING AND NOTHING THAT BINDS ALL TO THE SYSTEM OF THE MEGACOSM. ALL OF CREATION IS IN HIM, AND HE, IN TURN, IS CREATION. HIS ENTIRE EXISTENCE CAN BE SUMMED UP BY THE MIRACLE OF PHYSICS: THE ONE-IN-A-TRILLION CHANCE THAT THE UNIVERSE HAPPENS TO HAVE THE PERFECTLY-CALIBRATED PARAMETERS TO ALLOW US TO EXIST HERE AT ALL. ALMOST LIKE THE UNIVERSE WAS CREATED FOR US, WAS EXPECTING US. LIKE SOMEONE WILLED A UNIVERSE LIKE OURS INTO BEING, JUST FOR THE PURPOSE OF PUTTING US IN IT.

AND HERE WE EXIST, THE ONLY KNOWN POCKET OF LIFE IN 90 QUINTILLION LIGHT YEARS OF SPACE AND TIME. ALL ALONE, AS FAR AS THE EYE CAN SEE.

THE UNIVERSE HAS BEEN WAITING 13.8 BILLION YEARS FOR LIFE LIKE US. THE COSMOS HAS BECOME SELF-AWARE, BECAUSE WE ARE SELF-AWARE. HUMANITY CAN MAKE SENSE OF THE WORLD. AND WE CAN SAVE OURSELVES IF WE EMBRACE THE UNIQUENESS OF OUR DESIGN: THE BRAIN THAT HAS ALLOWED US TO BECOME THE MOST EFFICIENT PREDATORS EV-

ER SEEN. THERE'S STILL TIME TO END THE MASS MURDER. BECAUSE EARTH IS US, AND WE ARE EARTH. THE SAME SYSTEM, THE SAME BODY. A GIFT LIKE NO OTHER.

A BIRD NOW, AN ALBATROSS IN THE AIR. I AM AN EYE IN THE SKY, THE OMNIPOTENT BEING, GOD HIMSELF. I SEE IT ALL, BECAUSE I CAN SEE IT ALL. BECAUSE I HAVE EYES AND EARS, A NOSE AND A TONGUE, HANDS WITH THUMBS, AND A BILLION BEATS IN MY HEART. BECAUSE I HAVE A BRAIN, AND I KNOW HOW TO USE IT. BECAUSE I WAS MADE IN GOD'S IMAGE, AND GOD'S IMAGE IS DIVINE INDEED. I AM A SUPERHERO, A GOD AMONG GODS. I AM A HUMAN BEING, AND I CAN DO ANYTHING. BE ANYTHING. GO ANYWHERE.

THAT'S ME, COOPER, THAT LITTLE BOY DOWN THERE, SITTING ON ORANGE LIKE A VEGETABLE IN HIS PATCH. HIS ROOTS STRETCH TO INFINITY, THE TREE OF LIFE. ALL OF HUMANITY, ALL IN ONE PERSON.

THERE'S PANDO ON THE COUCH NEXT TO ME, LOOKING AFTER ME LIKE A BROTHER. HIS BODY FUSES TO MINE, UNIVERSAL BROTHERHOOD. ALL OF US BIRTHED FROM THE PRIMORDIAL SEED.

BUT ALL GREAT MEN ARE NOTHING COMPARED THE GREAT WOMEN WHO BIRTH THEM. NO SEED STARTS WITHOUT A WOMB. AND OUR WOMB IS OUR GREAT MOTHER EARTH, THE ONE WHO NURTURED US THROUGH THE LONG GESTATION. LITTLE COOPER DOWN THERE LOST HIS MOTHER TOO YOUNG. SHE MEANT THE WORLD TO ME. WE NEED TO BE SURE WE DON'T LET OUR GREAT MOTHER DIE BEFORE WE'RE TRULY READY TO LET HER GO.

COOPER SEES IT ALL NOW. I AM THE SEEDLING OF THE SEED, A MARVEL OF CREATION, THE BIRTH OF COSMOLOGICAL INTELLIGENCE. COOPER IS ONE OF THESE GIFTED FORMS, AND HE'S ALIVE AT THE UNIQUE MOMENT OF THEIR FINAL TEST,

THE MAKE-OR-BREAK THAT WILL DEFINE THEIR ULTIMATE DESTINY. WILL IT SURVIVE THE FILTER AND WIN THE UNIVERSE? OR WILL IT WITHER AND EXPIRE, MERE FOSSILS ON A FORGOTTEN WORLD, SPINNING ENDLESSLY IN THE EMPTY DARKNESS OF SPACE. THAT'S THE CHOICE WE FACE. AND COOPER CAN BE HUMANITY'S MESSIAH, THE SOLUTION TO THE PROBLEM. HE MUST FIND THE COURAGE TO STAND UP TO THE FACE OF EVIL. HE MUST BE ONE OF THOSE HEROES.

BUT HE'S JUST AN ANT ON A STONE. AND HIS WHOLE LIFE IS ETCHED IN SLATE. EVERY MOMENT OF HIS BEING IS LAIN OUT BEFORE HIM, TIMELESS, ETERNAL. HERE HE IS, A KID RUNNING THROUGH THE LAKESHORE WITH HIS MOTHER, SPLASHING WITH THE ELATION OF CHILDHOOD. AND HERE HE IS, A PRETEEN MASTURBATING FOR THE FIRST TIME, WONDERING WHETHER OR NOT HE HAS JUST COMMITTED A MORTAL SIN. AND HERE HE IS IN THAT WHIRLWIND THAT EXISES HIS MOTHER FROM HIS LATER DAYS.

HE WAS ALWAYS GOING TO LOSE HER ON THAT DAY. IT WAS DESIGNED TO HAPPEN. THAT'S HOW IT WAS PROGRAMMED. FROM THE FEMTOSECOND THE UNIVERSE BLINKED INTO EXISTENCE, THE ATOMS IN ROBIN BRAYDON'S BODY WERE TO RETURN TO THE UNIVERSE ON THAT DAY IN COOPER'S SIX-YEAR-OLD LIFE. WE TOOK HER BACK. SHE HAD FULFILLED HER PURPOSE, AND YOU, COOPER, ARE HER LEGACY. AND YOU WILL FULFILL YOUR PURPOSE, THE PURPOSE OF ALL IN YOUR COHORT: TO STEER THE SHIP TO SAFETY. BECAUSE YOU'RE GONNA USE YOUR BRAIN. YOU'RE SMART ENOUGH TO KNOW GOOD FROM BAD, RIGHT FROM WRONG, WHAT WORKS FROM WHAT DOESN'T.

THE CARBON ECONOMY DOESN'T WORK. IT IS A SYSTEM DESIGNED TO FAIL, IN A UNIVERSE DESIGNED FOR SUCCESS. WHY SQUANDER THIS ONE CHANCE? WE CAN FULFILL OUR

DESTINY, THE DESTINY OF CREATION. AND YOU'RE GOING TO DO IT, COOPER. *I'M* GOING TO DO IT. ALL OF ME. HUMANITY. I'M GOING TO SAVE THE WORLD, COOPER BRAYDON. HERE I AM, SAVING IT. HERE ARE THE YOUNG PEOPLE, LOCKING HANDS, SCREAMING IN UNISON, *WE'RE NOT GOING TO LET THIS HAPPEN.*

HERE WE ARE, STARTING OUR MOVEMENT. HERE WE ARE STANDING UP FOR OUR RIGHTS, AND THE RIGHTS OF EVERYTHING ON GOD'S GREEN EARTH. HERE WE ARE TAKING BACK OUR FUTURE.

AND HERE I AM, SITTING ON THE COUCH, TRIPPING ON ACID, THE WORLD AROUND ME ON THE CUSP OF RADICAL CHANGE. IT'S ALL BEEN BUILT INTO THE FLOORBOARDS OF TIME. I'LL TO DO WHAT I'LL ALWAYS HAVE DONE, AND ALWAYS WILL DESTINY. IT WILL BE FULFILLED.

HOW MANY PEOPLE HAVE TO DIE BEFORE HUMANITY WAKES UP TO THE HOMICIDE? WHEN WE'LL RECOGNIZE THE HOLOCAUST HAPPENING ON A UNIVERSAL SCALE? THIS IS YOUR CHOICE.

HERE YOU SIT, COOPER, ON THE PRECIPICE OF THAT CHOICE. AND HERE YOU ARE LEADING A MOVEMENT. AND HERE YOU ARE ON YOUR DEATHBED. AND HERE YOU ARE TRYING, AND FAILING, AND TRYING AGAIN, ALWAYS TRYING, NEVER GIVING UP, LIKE A TRUE ADULT. AND HERE YOU ARE IN YOUR DARKEST HOUR, WHEN FUTILITY HITS YOU LIKE A BRICK, WHEN YOU'RE QUESTIONING EVERYTHING IT MEANS TO BE AN ADULT AND A KID AND A HUMAN ALL AT THE SAME TIME, WHEN THE CLOCK IS ABOUT TO STRIKE MIDNIGHT, AND WHEN YOU'LL DECIDE WHETHER OR NOT THE INEXORABLE PULL OF GRAVITY WILL ROB YOU OF THE LIFE YOU COULD HAVE LED.

YOU CAN BE GREAT. YOU *ARE* GREAT. WHAT ARE YOU

WAITING FOR? SNAP TO IT. THE FUN IS OVER; GET TO WORK. CHILDHOOD ONLY COMES ONCE A LIFETIME, AND MOTHER'S ABOUT TO KICK YOU OUT OF THE HOUSE. DO YOU REALLY WANNA BE UNEMPLOYED *AND* HOMELESS?

SHOW MOM YOU CAN MAKE SOMETHING OUT OF YOUR LIFE. SHOW HER YOU KNOW HOW TO COOK A DECENT MEAL WITHOUT BURNING DOWN THE HOUSE. SHOW HER YOU CAN PAY YOUR DEBTS ON TIME. SHOW HER YOU CAN CLEAN UP AFTER YOURSELF. SHOW HER THAT YOU CAN GIVE HER GRANDCHILDREN ONE DAY, SO SHE WON'T HAVE TO WORRY WHETHER THEY'D BE BETTER OFF DEAD. SHOW HER THAT HER SACRIFICES WEREN'T WORTHLESS.

THIS IS ABOUT MORE THAN DOING THE RIGHT THING; IT'S ABOUT HAVING THE DIGNITY TO DO IT. THAT'S HOW YOU BE A MAN, A TRUE ADULT. AND YOU, COOPER, ARE ABOUT TO BECOME AN ADULT. ALL OF US ARE. WE WILL FLOWER INTO MATURITY, WITH THE DETERMINATION TO GET US THROUGH THE ANTHROPOCENE OKAY, TESTED BUT STRONGER FOR IT.

IF THE HOLOCENE IS CHILDHOOD AND THE ANTHROPOCENE ADULTHOOD, THEN HUMANITY IS GROWING UP IN MAY, WHEN THE LEADERS OF THE WORLD DECIDE WHAT KIND OF SHITHOLE WE'RE WILLING TO LIVE IN. THERE'S NO TURNING BACK NOW; 2C IS ALREADY HERE. BUT AT LEAST IT'S NOT 4C. ARE WE GONNA STACK THE DECK IN OUR FAVOR, OR ARE WE GONNA FOLD EARLY?

. . .

HE WALKS OVER TO THE WINDOW, UNFASTENING THE BOLT. HE SHOVES HIS HEAD CLEAR OUT, GAZING AT THE GROUND BELOW. FORTY FEET. INSTANT DEATH.

. . .

SHOULD HE JUST JUMP?

. . .

A collection of teachers mills about in the school library, chewing the fat over free cups of coffee. An assortment of donuts, bagels and cream cheeses is available on a table by a bookshelf, with a plate of fruits and vegetables beside it. Shree makes her way to this spread, grabbing a sprig of celery and chomping down in loud crunches.

Principal Sanchez raises her hand for attention. “Alright team,” she says, quelling the conversations. “Everybody take a seat.”

The meeting is taking place before school hours, and all of them are irritated from having to wake early. Shree is no exception; she takes a seat at the far end of the tables in the room, not looking to chat with anyone.

“We’re a month out from testing week,” Sanchez says, scanning the eyes in the room.

Shree tunes out the voice, losing focus as her thoughts wander. The principal begins to fire up the faculty, getting them amped for the month of dedication required for their students to succeed. But the pep talk means nothing to Shree; the cheating affair has delegitimized the entire institution. Now she knows that no matter what she does, the mistakes will be erased, the right answers scribbled in, the scandal covered up. Just a pack of lies, to complement all the other lies.

"Are there any questions?" the principal asks, looking around the table.

Shree's hand flies up.

"Yes, Sharifa," calls Principal Sanchez.

"So what happens if we lose our funding?" she asks.

All the eyes in the room shift to her uncomfortably, as if she had just violated a taboo.

Sanchez smiles politely. "We won't let that happen," she says.

Shree understands her meaning exactly: they'll just throw another bubbling party. She shuts her mouth until the meeting concludes.

Everyone rises to their feet, still plenty of time to hang out with each other before class. Shree picks up a few grapes from the table, then makes her way toward the back exit of the library, careful not to cross paths with Sanchez. Before she is able to depart, a teacher flags her down.

"Hey, Shree, did you ever look over that article I sent you about overcrowding in schools?" the elder instructor asks.

Shree shakes her head. "I've been really busy," she says, a weak but accurate excuse.

Two other teachers join her, followed by another two. "I read it," one of them breaks in.

"Yeah, thanks Judy," another says between bites of a donut. "I had no idea there would be such a glut in the years ahead."

"Are you guys talking about the population article?" asks another teacher, joining in the small crowd.

Shree watches as the women around her ignorantly discuss the challenges of teaching students in a world of overpopulation. More desks, more pencils, more paper, more wrong answers. And on the flip side, less floor space, less extracurriculars, less attention for each student. Oh what a challenge. Oh what fun.

"It's too bad we won't be able to feed them," Shree says, drawing everyone's attention her way. "Not with climate change destroying our crops."

The mouths in the room go silent; she said the C-words. All of them begin to judge her silently, wondering if she's some shade of retarded.

Her face flushes with hot blood. She shifts her eyes around, painfully aware of how ostracized she feels in this moment. It isn't that they don't know; it's that they know better than to talk about it openly. A bright red mist overtakes her senses.

"You're all *fucking* cowards!" she snarls, the denial too much to bear. She marches out the door, leaving her colleagues to resume eating their sugar pastries.

"Shree!" one of them calls. "Shree, wait."

She stops her fuming halfway down the hall, clenching her fists to face her pursuer. "What?" she asks her angrily.

"Are you okay?" asks her colleague.

"Do I look okay to you?" Shree boils. "Does any of this look okay?" She turns around and resumes her march, not looking back. She finds the nearest exit and slams through the crash bar. The April morning air caresses her face coolly.

It'll take an act of God to get these people to wake up, she thinks to herself as she walks down the sidewalk. When did *this* become the world she lived in? What happened to the world she knew growing up?

Though she is by all definitions an adult in age, body, and position in society, she knows she's as helpless as an infant. She feels her life becoming a fatalistic enterprise, her body jerked along by invisible strings, her path dictated by the angry carbon molecules in the air — tiny demons that have overpopulated the world. Her legs quiver as they carry her across the street, as if she's knowingly walking into the crematoriums at Auschwitz.

To die at the hand of carbon pollution is to die at the hand of herself, an act of suicide. For it is she that put that pollution there, she and everyone she's ever known. This reawakened carbon will enslave her race and everyone who comes after her to a hundred thousand years of suffering, a moral crime so unimaginable that she finds it hard just thinking about it, let alone coming to terms with it.

How could God do this to us? she wonders, looking at the naked trees with their budding leaves. *Why has He failed us?* The answer, to her, is obvious: we have failed Him. Whatever God had in store for her, it didn't look good.

She wanders to a park, children cavorting on the playground around her as she takes a seat. A kid of kindergarten age runs by, his ears pierced and a banana sticker on his forehead, so carefree, so unaware of what lies ahead. Another child is sitting on a bench, just crying, a terrible sobbing that activates Shree's maternal instincts. The wails give way to screams; there's no obvious cause.

Why is she screaming? Shree wonders. *Does she know?*

A small dinosaur walks up to her on the ground, a lovely pigeon with feathers of gray and iridescent green, bright orange eyes that stare at her with curiosity. She has never liked birds, but this time she pities the creature. It looks like it's begging for food. It should know by now that humans are nothing but danger.

Shree is a top predator, evolution's greatest weapon, for she and humans like her have hijacked command of nature, reshaped it, molded it, and created this new world, a world of our own design, colored by our own brush. We choose the species that survive, the plants and animals that serve us. The rest get cast aside, victims of a mass extinction they have no hope of surviving. And soon the tower of blocks will come crashing down, the ecosystems will unravel and the population will crash, taking her, and the pigeon, and everything else with it. In the meantime, however, she can sit fearlessly in the face of this small, wild creature, perched atop the food chain as the top killer on Earth, a mass murderer with a one-way ticket to Hell. But until Hell descends, almost nothing can challenge her supremacy. The only thing that can bring her down is her.

"I can't help you," she says to the pigeon sorrowfully. "If you think it's hard to find food now..." she trails off, not feeling the need to finish the thought.

The bird yaps and bows its head as if in understanding. It's the only nonhuman animal in the park.

Will pigeons survive? she wonders, looking into the creature's orange eye. It doesn't matter, she concludes. What matters is that they're sharing a moment together, two lives of equal stature, one unhappy, the other hungry, both of them hopeless. Murderer and victim, staring each other down. The poor thing knows it's hungry, but it doesn't know it's gonna die. Only one of them feels the weight of death, its bony hand reaching out to suffocate the world.

"Get out of here while you still can," she says to the theropod.

Realizing there's no food to be found, the bird flies away, leaving Shree to bask in her loneliness.

"Wait...don't go," she whispers. "Please don't go."

She looks back at the carefree humans. She preferred the bird.

She arrives home that afternoon, distraught and exhausted. And she isn't alone — on the couch in the living room, Mel appears to be in a state of misery. As Shree draws nearer, her roommate fails to even acknowledge her approach. Her cheeks are stained with tears, mascara in dry craggy streams like the Colorado River. A half-empty bottle of vodka stands on the table between the sofas, a single shot glass by its side. It looks like she's in the midst of drinking herself to death.

"Mel?" Shree asks, aghast at the sight. "What are you doing?"

Her roommate sniffs loudly, continuing to stare straight ahead. Her phone is in her hand. "Oh, you know, just hanging out," she says.

Shree takes a seat on the futon. "It doesn't look like you're just hanging out," she gasps. "It looks like you're trying to kill yourself."

Mel's eyes look over at her. "Well, Marquis is dead, so I'm just having a drink," she grumbles.

"What do you mean Marquis is dead?" Shree asks in a panic.

Mel burps quietly, almost vomiting in the process. "Car crash," she explains. Tears well in her eyes, which she casually blinks away.

"Oh my God," Shree exclaims. "I'm so sorry." She casts down her eyes, despondent from the tragic news.

"Yeah," says Mel, sniffling. "Stupid. He was taking a selfie from behind the wheel. It's the last thing he posted online. I think he lost control or something. Fucking idiot."

She grabs her phone and flicks on the screen, pulling up Marquis's online profile. His page has become a memorial, filled with posts from grieving friends and family as they relate their mutual devastation over his untimely death. She hands the phone to Shree, who scrolls through the heartfelt digital messages.

"This is so sad," Shree sighs, unaware that she could possibly feel worse than she already did.

"I know!" Mel howls, beginning to sob. She reaches for the bottle.

Shree stops her, pulling it away before she has a chance to pour. "No," she commands, returning the vodka to the table. "I don't know when you started, but this is when you stop."

Mel sniffles again, leaning back against the orange cushions. The mortality of life has never been clearer. "I just want to go home," she laments.

Shree moves over to sit beside her friend. "Hey," she says, putting her arm around her roommate. "You wanna go out tonight?" Mel answers with a sniffle and a smile.

And so they go, ignoring the hardships of adulthood. Two friends, bound for an East Village bar, ready to get wild and woolly on the dance floor, young adults in the prime of their meager lives. The club tonight is bumping, youthful faces in droves around them.

"To Marquis," says Mel, raising her glass to Shree's with a clink. "As he used to say, 'Beer rhymes with love.'" They chug their drinks in seconds.

Shree ahhs satisfyingly. "So you see anyone cute?" she asks, looking about in the dark club.

"Dude. So many." Mel points discreetly to a tall, blond-haired man in thick hipster glasses. "I'm feeling it," she says.

Shree slams her palm on the bar enthusiastically. "Alright. I'm your wing girl. Let's go."

The millennials join the dance floor, smooth undulations of bare flesh rubbing against the bodies in the room. It's pure sweat, adrenaline, sex. Minus Shree, they're all looking to bone, dancing to attract themselves a mate.

Mel asserts herself into the tall man's group, showing off her body in a welcoming series of movements. She shakes her ass briskly in his direction, receptive to an advance. He takes the opportunity and they grind in harmony, one wild girl, and another wild guy. They're both drunk, they're both free, and they both don't care; they're just having fun, like young people should.

Shree watches in satisfaction as Mel enjoys herself. But nature calls, and she gravitates toward the bathroom, standing in a long line pressed against the wall, each one of them waiting to relieve themselves. The dancers encroach upon their space uncomfortably, occasionally bumping against her without so much as an apology. There's never any space in New York, no matter where you are.

After a few painful minutes of holding it in, Shree reaches the front of the line, and soon enters the solitary stall. With the urgency of labor, she runs inside as soon as the door allows, locks the knob, and pulls down her pants as she squats on the dirty porcelain bowl. That it's wet under her cheeks isn't important to her; the sweet relief of urination is all that matters. She can worry about hygiene after her bladder is empty.

The toilet flushes with a raucous splash, another gallon of wastewater for the New York City sewers.

Shree's guilt returns as washes her hands. She can't help that she's a human that needs to pee. And she can't help that the only place to do it is into a small well of freshwater. But she *can* feel bad about it, and maybe that's a start.

She turns off the tap and dries her hands with a paper towel. But the faucet catches her eye: a small stream of water trickles forth from its spout. Annoyed, she tightens the knob as far as she can. The water continues to flow.

She turns the water on again, only to turn it off more forcefully this time; no use. *What the hell, faucet,* she mulls angrily. She tries again, harder. Still no dice. She begins wrenching the faucet head violently, attempting to stem the flow.

"For the love of God, please turn off!" she shouts as she strangles it with her grip.

The rusty end of the spout detaches, and the trickle gushes into a deluge. "Fuck," she curses to herself. "Fuck, fuck, fuck." Water pours freely from the tap, emptying straight down the hole at the bottom of the basin. She tries to put the cap back onto the head of the spout, but it comes loose under the force of the water. It's broken, stuck in perpetual full stream.

"God dammit!" she shouts in frustration, chucking the rusted knob of metal onto the sticky tiles. She holds her head in shame as she exits, just another sinner wasting the future away. This is the worst day ever.

Helplessness Blues

Orders are shouted like roll call in the barracks; under-caffeinated Americans shuffle into the coffee shop looking for their daily drug. Cooper's Straunt uniform, though clean, has faded considerably since he first put it on. He follows his habitual patterns of execution, mindlessly filling Styrofoam cups with oil-black liquid, adding the requisite sugars and creams specified on the ticket labels.

The joy of being employed has long since worn off. If going to work every day for the rest of his life means coming to an experience like this — a place he doesn't care for, doing things he doesn't care about — then it's going to be a long slog to retirement. Despite his good intentions, he feels he will never amount to anything more than a chipped tooth on a tiny gear in some insignificant corner of the global carbon economy.

At the first available moment, he steals himself away from the madness of the kitchen and collects himself for a moment in the employee restroom. The morning rush has died, the chaos ceasing at least until lunch. No better time than now for a quick smoke break.

Cooper pulls out a plastic baggie with a small, pre-rolled joint tucked inside, along with a royal blue lighter, smudged with ash. He grabs them both in his fist and lights up, puffing gently in the employee bathroom. His brain melts into a putty, calm but alert. Within a minute, he's baked.

He snuffs out the smoldering end of the joint by pressing it into a dusty aluminum basin. He returns the roll to the bag, careful to save it for as long as it can last; drugs aren't exactly high on the priority list when there are bills to pay and mouths to feed, namely his own. He flushes the toilet, then sprays two long streams of air freshener into the cramped lavatory. He opens the door to return to shift.

"Mmm," says Nilas as he watches Cooper exit. "Something smells good."

The boy smiles guiltily and turns back for the kitchen.

"I didn't know you smoke," Nilas says as he walks by.

"Everybody smokes, man," he replies. Between his worsening grades and being targeted by Cokeheads, a quick puff of weed is all he can do to have any control over the world, even if it's just over himself. It's the only thing he has to look forward to. After this, it's just more of the same — the long, slow descent into Hell.

At this point that it dawns on him that he's living someone else's dream. He shows up to work so he can afford the food and drink that will keep his cells alive a little longer, but for what, ultimately? So he can pass on his genes to another generation of doomed humans? So he can watch stupid internet videos when he gets home and post flattering self-portraits on social media? So he can get the degree he doesn't want, with the money he doesn't have, so he can participate in a wasteful world he doesn't agree with? None of these motivations are strong enough for him. This isn't the life he wants at all.

He returns to his spot behind the counter, ready to serve the next customer, and the one after that, ad infinitum. And as he resumes mindlessly pouring coffee into cups, he can't help but think of the overinflated value humans tend to place on themselves, their self-aggrandized egos steering them in the wrong direction. Why can't he be like everyone else his age, taking pictures of himself every four minutes and putting them online? Why can't he play the game right? Why can't he *matter*?

In the grand scheme of things, with the glaciers melting and the ocean boiling, his minimum wage job feels very small indeed. The act of keeping himself alive seems almost pointless; his future doesn't seem that exciting after all.

But this, he knows, is ultimately a mindset of defeat. He has shed any vestige of idealism from his disposition toward life, hardening himself to the hostile environment he now inhabits. His world has shrunk, and with it his ego. He acknowledges the one incontrovertible fact that no matter how strong his will, he alone can never get done the things that must be done. And at times when it feels like he's one of the few people actually making a full effort, it's hard to escape the interminable loneliness that surrounds him like razor wire. He wants to save the world, but everyone else is playing a different game.

The result doesn't have to be sorrow, he knows. Maybe, instead of seeing the big picture, the key to sanity is in seeing the small. Next week is the big climate conference, the last hurrah, one of the final turning points in the history of civilization. It's either a beginning or a finale, and all signs to date have read "*Dead End.*" But Cooper can't do anything to influence the outcome of that meeting, and he knows it. He's but one person, a single voice. In this digitally-distracted world, only a chorus can be heard.

Still, he does have agency over some parts of the picture, no matter how small. And in this moment, he figures, his mark will be in making someone's day with excellent service. The world may be ending around him, but dammit at least he can be a good barista. It's all he has.

A homeless woman walks into the store. Her face is thoroughly wrinkled, her gray eyebrows like bushes above the dark slits of her eyes. She hobbles in on a cane, pulling out two folded bills from her torn jeans pocket. She mumbles to herself quietly as she approaches Cooper.

"How can I help you, ma'am?" he asks her politely, not judging her for who and what she is, who and what she has become.

"Double of espresso," she spits, handing the faded notes.

With a couple taps of the touchscreen, Cooper opens his drawer and drops them in. A long tongue of paper prints, which he gives to the woman, as if she needs it.

It's clear her plight has been much harder than his, and he's only too aware of it. Who knows how many bridges she's slept under, how many aches and pains she must have, diseases and rashes, all the nasty things that come from poor hygiene and malnutrition. It's a miracle she even has the patience to deal with it.

In 21 years, Cooper has never experienced true pain, like the kind most children in history have faced as they grew up. As far as bodily harm goes, he's only ever broken his toe. He's never had major surgery. He's never been abused, never tortured. Until this year, he had never felt starvation, and even then he's still functioning. He's always had a home over his head. He's had running water and electric light, the latest digital accessories, and constant streams of entertainment. He's lived the cushiest existence anyone could live.

And here's this person, across the counter, who has nothing, and probably always will.

We'll all be homeless soon enough, he reminds himself. In this way, he doesn't see the woman as a stranger; he sees her as himself. A reflection of the future: poor, starving, and homeless.

On the giant screen before Pando is a video feed from an unmanned drone. A smaller display to his right shows detailed information of flight conditions, aircraft status, and map coordinates. Iri watches silently from behind.

"There it is," she says upon sighting the familiar landmark. The gray building on the screen brings back painful memories of the first shots of the climate wars, the hideous murder of her friend. Payback is so close.

"I see it, I see it," Pando assures, his scintillating eyes centered on the screen. "El Chupacabra."

Behind the controls of this wasp in the sky, he maneuvers the hovering craft closer to the installation. But as it approaches, something feels off. The building looks different.

"What's going on?" Iri wonders, high panic setting in. Pando brings the plane in closer, only half a mile away now. He starts descending toward it.

The roof of the warehouse bay appears to have opened to the sky. The canisters of aerosol that sat there waiting for deployment are nowhere to be found; it's just an empty space. Moreover, the building appears to be in a state of deconstruction, being dismantled by a nearby crew of construction workers. The junkies are trying to cover their tracks.

"Switch to infrared," says Iri. Pando obeys, the colors on the screen mutating from greens and grays to yellows and reds.

No fewer than twelve moving bodies can be seen through the heat lens, working like termites to take down the building. One of them seems to have noticed the drone humming in the sky.

"I think we've been spotted," says Pando. "Should I pull up?"

Iri shakes her head. "No, get closer. I want a better look inside."

The drone is now whirring directly over the building, its camera examining everything in sight. But there's nothing to see. Just an empty building.

"They're gone," says Pando. Goosebumps pepper his skin like a basketball.

Suddenly a bright flash appears from the bottom corner of the screen. It fumes from the back like an RPG, racing into the camera before exploding into a fireball. The screen cuts to static.

Pando inhales sharply and holds his breath. He turns to Iri, who is frozen in shock at what she has seen on the screen. She closes her eyes slowly, drinking it in. At last Pando exhales, his hands dropping from the controls and into his lap.

The geobomb is missing.

. . .

"It's time to up the ante," suggests an ominous voice. It is but one of many to be heard around the table, each voice catapulted into near-shouts by the recent turn of events.

At the sound of his words, the heads in the room snap to his attention. They were all thinking it, but no one knew how to broach it. True to form, the delegate from LA is honest enough to admit when the risk is too heavy to bear.

"You're talking violence," says Nevaeh, reading Santi's face. The Californian nods quietly. The time has come.

"I agree," says another voice from the far side of the round table. The Clean Front delegates have been called to an emergency meeting on this warm April night, here to discuss the latest. More representatives chime in until the vast majority is in favor. Armed insurrection seems to be the only course of action that remains.

"How?" asks Val, still wearing her trademark Phillies hat.

Santi shrugs. "I don't know, but we need to get people's attention *right now*, and violence tends to do that," he says. "We have precious few options left."

"We could bomb a Coke building," suggests a Chicago delegate.

"No," Nevaeh halts. "We need people to sympathize with us."

The leaders ponder the event that will catapult the clean revolution into public view. Whatever it is, it will have to be the perfect catalyst; you only get one shot to start a revolution.

"Why don't we just publicly announce that the Cokeheads have a geobomb?" asks a representative from Galveston.

Santi shakes his head. "The public doesn't respond to climate change with any interest," he points out. "Even if we did reveal what we know, the media would suppress it immediately, and the Cokeheads would probably sue us all to death."

"Which is *why* we need to elicit sympathy from the public," Nevaeh repeats. "Are you not seeing this?" They consider her curiously; none of them know what she's driving at. She looks around at the twelve pairs of eyes in the room.

"We're going to bomb a protest," she states.

Santi blinks confusedly.

"An anti-green protest?" Val asks.

"No," says Neveah. "The Columbia divestment protest."

The eyes stare back at her, dumbstruck. They wonder whether they heard her correctly, and, more importantly, if she meant it.

"You can't be serious," says Santi.

Her eyes glint revealingly. "But we frame the Cokeheads."

They see she's onto something.

She smiles as she leans into the table. "Think about it," she explains. "Students are killed protesting fossil fuels by renegade carbon junkies working for the Coke Coalition. The public is outraged. Counter protests are staged throughout our territories, at every university in our cities, in solidarity with our fellow students and citizens."

She lowers her voice, adding flair to the captivating delivery that has everyone's eyes locked on hers.

"Then," she resumes, "when the public is boiling with rage that our students' lives are being taken by carbon polluters, we reveal the existence of the geobomb, and we sink the Cokeheads once and for all."

Santi nods vigorously. "And then we track down Coke and lead a mission to destroy that geobomb, no matter the cost," he says.

Nevaeh nods at him understandingly. "Revolution begins. Simple as that."

Iri scoffs. "It's never as simple as that," she remarks. "But it's that or nothing."

"Can we even pull that off?" asks the leader from New Orleans. "How are we gonna frame them convincingly?"

Nevaeh shakes her head. "Leave that to BKP," she assures him.

Val shakes her head, disturbed. "We're talking about innocent lives here, Nev," she whispers to the group. "Risking *our* lives is one thing, but risking civilians—"

Nevaeh raises a flat hand, ceasing Val's thought. "I make the suggestion lightly," she says. "But if we're at the point where we have to use violence, we have to frame the Cokeheads. It's the only way to turn the public against them."

"I'm with her," says Santi. "It doesn't have to be anything big. A few casualties would be enough."

Val nods thoughtlessly. "Alright, fine," she says. "Maybe it will work."

Not one person in the room is a hundred percent sold on the plan. But with a geobomb potentially armed somewhere in the world, the stakes are tens of billions of lives — as they always have been. The math makes sense: two to three lives today for the sum of unborn humanity of tomorrow. Sometimes there is no good option. But it's better than no option.

"Then we're gonna do this," Nevaeh says, once again leading the pack through the darkness. "BKP will coordinate the bombing, while the rest of us will rally around this tragedy. We need to have posters and brochures ready to go. The minute the bombing breaks on television, we hit social media and we start to protest. Is that clear?"

She holds out her hands to the delegates beside her. They each grasp the warm flesh, and in turn extend their own hands down the table, forming a complete, unbroken chain.

"We are in unity on this," she affirms. "As long as we're together, we can't fail. And we won't."

The geobomb is missing. The geobomb is missing. It's *missing.*

Pando can't shake the thought from his head. Where could it be? On a plane? Already deployed? He perspires alone in the smoky room, the thick scent of bud heavy in the air. With nothing known and everything at stake, the refugee takes hit after hit of the green drug, shrugging off the weight of the world until he gets new orders regarding this disturbing new development.

It's missing, he thinks again; the weed isn't tuning out the threat. He replaces his pipe into the yellow plastic bag, and sits in solitude for minutes, hours, doesn't matter. Every passing second is a moment closer to doomsday. But doomsday may have already arrived.

Through the front door arrives Cooper, surprised to see Pando home so early. It's been some weeks since he hasn't put in a long day at BKP.

"What's the news?" the boy asks. He knows the drone mission was carried out today, and he's been dreading the outcome ever since his friend signed up for it. He really didn't expect to be a soldier in a war like this. Most people never do.

Pando is locked in a thousand-yard stare. "Oh, you know," he mumbles. "Another heat wave in India; this time it only killed a thousand people." He continues to peer ahead blankly. "The Islamists are beheading Americans. Black teenager shot dead by police — you know, the usual."

Cooper rolls his eyes. "You know what I mean. What's the *other* news? How did it go?" Pando doesn't look at him, instead staring directly ahead.

"Not good, huh," Cooper understands.

Silver eyes slowly turn to face him. "What do you think?" he asks pointedly. "Is anything ever *good* anymore? *Huh?*"

Cooper winces. "I guess you'll tell me about it later, then," he says.

Pando relaxes on the couch again. "I'm sorry. I'm just..." he fades, shaking his head. "Get ready for the Anthropocene."

The adolescent nods uneasily, a shiver crawling up his fragile spine. Anthropocene. The geological age bearing the mark of man, when humanity's presence becomes the dominant driver of the systems of Earth. An age of man, or, specifically, "new human." In this case, new probably means extinct.

"I need to study," Cooper says, excusing Pando. "My grades have been going down the shitter."

He leaves Pando in silence as he climbs up to the loft. In his room, he pulls out his laptop and a brick textbook from his bag. He opens his browser instinctively, his brain seeking instant stimulation. He knows he should be focusing, but his focus is on autopilot, drawn to the internet like a moth to flame. His homepage is the BKP website; its leading headline reads "*Senator Coke Avows His Party Will Oppose Any International Climate Treaty.*"

News is never good. He closes his browser and buzzes Natasha instead, asking her what she's up to. He hasn't heard from her in nearly a month.

Natasha: Planning. Divestment protest next week. Come. 4:57pm
Cooper: I'll be there 5:00pm

He tosses his phone onto his bed and takes a look around his pitifully small room. In seven months, he's hardly taken any time to clean the miniature space he inhabits.

He grabs a dustpan and broom from downstairs. He gives the floor around his bed a quick sweep, picking up the loose hair and flecks of dead skin that accumulate in any space frequented by living bodies. The bristles swish under the table, carrying a host of debris, everything from crumbs of cereal to a stray rubber band. He brushes the pile into the dustpan and carries it down to the trash bin. It tumbles into the black plastic like a landslide.

As Cooper looks down at the mess he just deposited, he considers the rubber band at the top of the pile, pausing to examine it closely. Such a curious object; a stretchy ring. Who knows how it was made, where it came from, how many people and plants and energy it took to get this to him. Only for him to throw it away.

He plucks it out of the garbage and slips it into his pocket. It's still useful, after all. It isn't trash yet.

Mel arrives home from work not long afterward. Upon entering the apartment, she sees Pando nearly comatose on the couch, a vegetable incapacitated by marijuana, completely disengaged and dissociated from the world around him. She stands directly in front of his face.

"Hey, it's your turn to clean the kitchen and living room," she tells him. "I expect it to be done by the end of the day."

Pando's stare moves up to her. "Sure, if you clean the bathroom."

Her face protests. "No way, it's Cooper's turn to do that."

"Says who?" he asks. "We never made a cleaning schedule. You're just making this shit up as you go along."

Her eyebrows slant angrily. "No, *I'm* just trying to clean this apartment," she says pointedly, "unlike you, who's just sitting on your ass as always."

He blinks slowly, not breaking his vicious eye contact. "Well maybe if you bought soap once in a while, you'd actually manage to get some cleaning done," he says with calm aggression.

"At least I don't pee in the shower," she says, turning toward the ladder. "And at least I'm not fucking gross."

Pando's face turns bright red; the months of simmering frustration are erupting into uninhibited rage. His patience has run out, the spark reaching the end of the fuse. Combustion.

"Ya know what Mel, I'm getting real tired of your attitude," he sounds, leaping from the couch. She turns around suddenly in surprise. Pando has never been this direct with her.

"You know how much water I save by peeing in the shower?" he lambasts. "And you want me to stop just because you think it's gross, even though it has no effect on you whatsoever? How selfish can you be, Mel?" He raises his voice to a roar. "The world doesn't give two shits about what you think is or isn't gross, and neither do I. But you don't understand that, do you?"

Mel has no words. She just stares, awestruck.

"No, you don't," he huffs. "You have no *fucking* clue. So shut your goddamn mouth about where I choose to urinate. I've got bigger things to worry about than your bitching." And with that, he storms out of the apartment.

By the time Shree gets home from school later in the afternoon, all she wants is to relax in bed with Pando. She hikes to her room and closes the door, but as usual, her man isn't home. She sighs in frustration from another disappointing day at work. Before she can take off her bag, she hears a knock.

"How are you feeling?" Mel asks.

"Broken," she replies, letting her in.

"Do you wanna talk about it?"

Shree shakes her head. "I dunno."

"Talk to me, Shree," Mel says, taking a squat on the bed.

"The principal gave me a performance review today, and she says my job is in danger," Shree grumbles. She runs her fingers through her black hair, fallowing it into rows. "I feel like I'm not worth anything."

"You're worth a lot to me," assures Mel. "I think I'd go crazy if you weren't my roommate."

Shree smiles warmly, her soul lifting ever so slightly. "I'm glad you're here too," she says. "Pando's been so busy that I rarely see him. But you're still here for me, like always."

"Fuck him," says Mel. "I know you have feelings for him, but like, seriously, fuck him."

The teacher sighs with exasperation. "I'm not even sure if I have feelings for him anymore," she utters. "I don't have feelings for anything anymore." The weight of reality is crushing her like an anvil.

"Hey," says Mel, putting her finger under Shree's chin tenderly. "I was thinking of going to Cape Cod tomorrow to visit my mother and her boyfriend. Things have just been so crazy lately. You wanna come with me?"

Shree shakes her head. "I gotta get ready for testing."

"You can get ready on the beach," Mel implores. "You need a break. We all do."

The teacher considers it uneasily. It is a tempting offer.

"Come on, Shree," Mel pleads. "We need to get out of this filthy cockroach nest. It's killing us."

Shree finds herself warming to the idea. "Well, I could definitely use a vacation," she admits.

"Take Monday and Tuesday off," Mel says. "Let's blow this clambake."

Shree likes the sound of this, because fuck school and the liars she works with. They can cheat without her for a couple days.

"Alright, sure," she agrees. "Why not."

Solastalgia

Like a tornado coming to a sudden halt, Pando awakes from his nightmare in a cold, blinking sweat. Shree isn't really dead. Cooper hasn't blown the mission. The Cokeheads aren't torturing them in Guantanamo Bay. It was all just a dream.

He was back in Kabubwason, trying to save his brother, only to watch helplessly as he struggled in the water. But they were actually in New York, and his brother had been replaced by Cooper. Manhattan was taking on water, and on the news was a swirling demon of yellow, red and purple, sliding menacingly into New York Harbor. Hurricane Sandy II, according to the on-screen graphic. The world wars in his dream were a thing of the past; nations didn't cooperate like that anymore. The only thing they fought were the onslaught of city-destroying storms.

Just a nightmare.

With a heavy sigh, he rolls onto his side, wrapping around — oh wait, Shree's in Cape Cod with Mel. His arms collapse into the empty space, ruffling the bedsheet underneath. Sad to be alone, he stares at the ceiling, gently rubbing his beady forehead.

The arctic air of winter has fled the apartment, replaced by the sultry heat of summer. It adds a deadened layer of weight to the atmosphere, turning everything a hazy sepia, inside and out. Winter gave way to summer like the flip of a switch.

His body lies like a felled oak, bound to obey the law of inertia. His tan frame rests bare in the bed, his supple muscles brimming with untold strength, though currently sitting dormant against the mattress. They sit like chained pit bulls, ready to be unleashed at their master's command. Pando, in his current form, could probably kill any man his height with his bare hands.

He spies his dumbphone on the floor by his arm; he types out a message to Shree:

Pando: Have fun out there ☺ 11:23am

Shree gets the message in the middle of posing for a selfie with Mel and Cooper as they sit beside one another in the jazzy felt seats of the bus. She uploads the snapshot to her feed, where she tags her roommates so they can interact with it from their phones. Cooper Likes the selfie with smug irony. Mel and Shree have beat him to it.

Sitting on the upper level of the bus has afforded them the best views of the city. Cooper marvels at the thrilling New York morning, milky sunshine caressing every corner. Mirrors tower in the sky, bouncing light in all directions. And so many people, whirring along beautifully on the streets below. These were the days, man. These were the days of civilization.

But these streets aren't ours anymore. They've been conquered by that most indefatigable of conquerors, the ocean. These streets belong now to Neptune, who will relinquish them someday far in the future, long after mankind learns the lessons of this sinful era. But Cooper's legacy will outlast that, and so will the memory of this city. As long as he's alive to fight for it, he will make sure the world remembers New York. Remembers that once upon a time, there stood a metropolis on these shores, and it was one of the holiest sights human eyes could ever see. A city of light and wonder and strength and possibility. The future, as seen from the past.

As marvelous as the glinting obelisks are, Cooper sees through their hollow promises, straight to the rotten core of these phony polished teeth.

Every single one of them is an enemy, each one emitting a cloud of deadly gas into the air, drawing into rivers and eddies, whirlpools and rip currents, draining into the sky like sewage into the sea. And as long as they're powered by carbon, they send the Keeling Curve surging higher, a staircase winding skywards without limit.

The bus passes along the straight avenues, Cooper scrutinizing the edifices as they go by. Central Park looks as a jungle from Columbus Circle; Graham Court, with its palatial sophistication; another building of stone, sporting four massive pillars, the fat fingers of justice; the same chain of cheap fried chicken that he frequents when he has a couple dollars to spare.

He glances back at the buildings. They are false idols.

Just passed 133rd Street; he's seen a hundred blocks of this city since embarking. These were the streets, man. These were the streets that made history, the streets that raised the toughest, most adaptable people in the world — New Yorkers. Here, you could be anything, anyone. You could meet anyone, go anywhere. These streets incubated the poets and artists, streets that were graced by the kings of the world, streets where, even if you were a nobody, you always felt you could be somebody, in a country that allowed you to be somebody. These were the streets.

A flag at half-staff draws Cooper's attention to a sunlit school from the advent of the 20th century, children playing in a golden patch of pavement as the haze diffuses the sky into a homogenous white. They're baking in the sweltering warmth of the second heat wave this year; May has only just started. Still waiting on those April showers.

Crossing into Jersey now, the railroads beneath slice through the land along its coast. Rusted bridges as far as the eye can see, from the really exciting days of industrialism. The days when people sang *I Been Workin' on the Railroad* and meant it. The days when progress seemed limitless, as long as we kept burning oil. A false idol.

At this point Cooper realizes the carbon economy is ultimately something of a bait and switch. The fossil fuel industry promised us cheap energy; we bought it. In exchange we got a death sentence. Kind of like buying a cheap car, only to make it to the highway and realize the brakes don't work.

If they had told us this cheap energy came at the cost of 90% of life on Earth, we probably would've called it too expensive. Most people would rather be responsible and buy a better car with working brakes. This planet of ours is kinda like a family car: we want the kids in the backseat to have a future.

As he looks back on Atlantis receding in the distance, he thinks about all the contradictions, all the bullshit we use to justify our actions. Maybe it's money. Maybe it's God. It takes a dedicated mind to wade through the deep pools of truth, and to shatter the thick crust of digital distraction that assaults the senses at all times. Ultimately, it all adds up to this: if you care, you must act. It's as simple as that.

His stomach quivers into a knot at the sight of the buildings; they make him sick. And the millions scurrying about in their crystal spires in the distance, acting like they care but don't, are just another symptom of society's insanity. It's madness. It's schizophrenic. It's suicidal. And it's human.

Can animals commit suicide? Cooper wonders. Probably not, he reasons. Suicide requires self-awareness and an abstraction of the future like only humans possess. Suicide may be the bottom line, the sole defining trait of humanity. A tendency toward self-destruction.

There is a malodorous stench in the air, emanating from someone in an adjacent seat. Maybe it's Cooper himself. It is, after all, a sweaty day; the mercury is expected to reach 88 today. To think it was only in the mid-fifties a week and a half ago.

The bus drives clumsily over the bumpy asphalt, rocking him in his chair. They pass an abandoned construction project along the highway; nothing remains but ripped up ground and sagging fences. Crooked signs spear the dirt at odd angles, leaning like drunken trees along the road. The eastbound traffic is a clogged artery, tiny tails of carbon coughing out the vehicle behinds, all of them heading to work for the carbon economy. Tick, tick.

. . .

In a stately shoreside manor, Mel sips a raspberry smoothie as she sits on the hardwood floor. Her mother sways back and forth in the rocking chair with the grace of a swan. The two share a tender moment in this well-appointed living room.

"It's a nice place," Mel utters, glancing around the room that used to belong to her grandmother. Nice doesn't even cut it; this is luxury. The mahogany bookshelves, the hand-stitched upholstery, the unused china on display. It almost feels like too much.

"I thought so," says her mother. She takes a hard look out the broad window to her left. "A nice place to retire."

Retirement. The thought doesn't sit comfortably with Mel. Only old people retire. Is her mom really that old already?

"You've got a long time before you have to worry about that," Mel says, hoping to reassure her mother. After all, she's only been working for forty years.

The elder woman smiles quietly, not turning her eyes from the shore. "I'm getting old," she says hoarsely. "I can feel it."

Her daughter's eyes flare with fear. "You're not old," she insists. But she can see it — the lines on her mother's face, the streaks of gray in her fraying hair. The eyes that have seen as much as anyone, like Florence Owens Thompson looking over the plains of the Dust Bowl. Her mother *is* old.

"It isn't just me," the woman continues. "Things feel old, too. There were birds here once, they used to sing. There are no more songs anymore."

Her daughter shifts uncomfortably on the floor. The mother peers down at her offspring.

"You're too young to understand," she says.

Mel does understand, though she never paid enough attention to the local songbirds to notice their absence. But the area does seem to have changed a bit over the years, dewilded somewhat. Whereas once it felt like there were feral animals in the woods, it now feels like there are only humans to be found for miles around. All others have been pushed out.

Mel's generation will be the last link to this world that was, the last living bridge between the old and the new. Grandchildren will hear the stories about this era, about the creatures that roamed the land. In time, they will pass from living memory. But they will never be forgotten.

After a quick breakfast, Mel and her friends are out the door, destination beach. The air outside steams with humidity. In nothing but bathing suits and bikinis, the three of them define the shape of blooming adulthood — tall, developed, educated. They trod the sand as young gods, unfurling their colored blankets like parachutes.

Mel's thoughts churn, digesting the possibility that her mother has entered the dominion of old age. A frown covers her face, broken only when Shree holds up a phone for a selfie. After the flash, Mel's smile quickly reverts to a frown. Enjoying reality is now limited to posing for pictures. As long as the internet thinks she's happy, does it matter that she really isn't?

Shree pushes buttons on her screen and posts the image into the digital ether. It amasses Likes, and even inspires a comment or two.

Mel's crisis continues to rattle her mind. If her mother is old, does it mean she's on the way out? Is her expiration date around the corner?

This is all too much to think about right now; she's supposed to be enjoying the beach. Not far along the shore, a group of people are playing volleyball. It's hard to tell from this distance, but at least one of them could potentially be hot. The game's afoot.

She pulls out a bag of makeup and begins applying it to her face, looking on from her front-facing phone camera. Just a few brushstrokes are all it takes. She looks at the blonde girl on the screen; damn. What guy would say no to that?

"Wanna take a walk, Shree?" she asks.

The teacher looks over to the volleyballers in the background. "Not right now," she shrugs, reclining into her towel.

Mel frowns deeper. Cooper has already put in earbuds and is tanning in the sunlight, eager to shed his white complexion. She decides to head over alone.

Shree feels her stomach groan sharply as she thinks about what's gonna happen to her if her students fail. So many students; so many tests. With a class as large as hers, disaster is the only outcome.

How could God put her through this shit? Maybe it's her punishment for being such a sinner, not only failing the future as a carbon polluter, but also as a substandard teacher. Despite all the logic against it, she feels like she's the problem. On the outside she may look relatively stable, but underneath the strained laughter and empty reassurances is a soul ready to break. How long she'll last at this point is anyone's guess. Anyone but hers.

Cooper pulls the plugs out of his ears. "I'll be right back," he says, motioning to the bushes in a thicket of nearby wood. He grabs a bag of drugs out of his backpack and lights himself a bowl among the trees.

When he returns to the beach, he doesn't sit down. The slapping waves against the mud seem to be beckoning him their way. Despite his inability to swim, Cooper obliges to the call of the sea, wading up to his waist without thinking twice.

The water is cold as frost on his body, but in the blistering heat of midday, it's an immaculate sensation. The grass growing in his brain warps his perception with each further step. By the time he's neck deep in saltwater, the land and sea appear as neon greens and blues through his eyes, which blaze redder than 22nd century Earth. A radiating deity surges overhead, a white disc against the electric sky. He paddles toward it, his feet lifting off the floor of the beach, flailing into open ocean.

He wants to gain its energy; he feels it flowing through him. On a conceptual level, he knows the sun is what feeds him. A ray of sunlight is turned into food in some autotroph somewhere, travels up the food web, and leads to him, perched right at the top. Sunlight flows to him from every source, in every form.

The sea is a bath which cleanses his spirit, returning his soul to the primordial spark that birthed life some 3.8 billion years prior. Trees dot the shore all around. The breeze whips past his ears, blowing the current against his path at an oblique angle, relentless wave after marching wave. But the light draws him nearer. He presses on toward the shimmering beast in the sky, eager to bask in its energy, to touch it, to become one with it. The power it holds over him is resolute, pulling him along further from shore.

He realizes his feet no longer have a connection to the ground. His mind panics, thinks he's about to drown without the safety of land. But he doesn't. His will tames the irrational thoughts, and he tells himself not to worry. He calms down and turns back to the shore, paddling like a dog. Within moments, he regains his footing.

He glances back at the star in the sky. In some respects, the Romans were right in worshipping gods of Sun, Moon, Earth, and the like. If humans are the product of divinity, then our natural creators are these celestial bodies. We're moss growing on the skin of a rock spinning blindly through space. We're the product of the savannah that led our furry ancestors down from the trees and onto the ground, where we stood upright and grew intelligence. We're the product of war, as the scarcity of resources led us to develop tools and weapons to out-compete our less clever contemporaries. We're the product of all the things the ancient civilizations worshipped.

Cooper begins to see his origins through scientist eyes. His thoughts reach back eons to the fiery origins of the universe. He sees the first stars, the first planets, the first supernova, the first seeds of life. They coalesce into the solar system, with Earth its crown jewel. He sees its first drops of water, its first mountains, and its eternal dance with its companion moon, around and around their holy body at the center of it all — the Sun, a brimming star in the prime of its life. Each star beyond is its own flaring deity, a giver of life to its offspring planets. We just happen to be one of the lucky few to have a home right where we need to be.

He turns back to the shore, swimming like a fish, the first vertebrate to evolve on this planet. When the water is shallow enough, he begins to crawl along the bottom like an alligator, four on the floor. He experiences the transition from sea to land as many different species from all branches of life, from tadpole to turtle. Finally he crawls out of the soup, emerging not just as a mammal, but an upright human. He reclaims his spot on the beach.

Hours pass, and Shree finds herself feeling no better. She tries to read a book, but she can hardly concentrate on the words. She takes the occasional swim to relieve her always-full bladder; "I gotta pee," she'll announce. Even her music doesn't seem to help. The environment around her may be idyllic, but it does little to soothe the demons swirling tumultuously in her head. Maybe she *is* worthless. Maybe she shouldn't be here.

Mel returns to sit beside her, the mission having ended in defeat. She didn't get anybody's number, and none of the attractive guys expressed any interest. Neither, for that matter, did the girls.

"You need to get rid of Pando," Mel suggests, reading the sorrow in her friend. "You weren't like this before you met him. You were much happier."

Shree rolls her eyes. "I was young then; I didn't know any better." She glances at Mel with sheer pain on her face.

"You still *are* young," Mel says, echoing her previous conversation.

"I don't feel young," Shree laments. "And it's not Pando's fault for helping me see the world as it really is."

"You mean how he wants you to see it."

"No, Mel, for what it is," Shree says sternly. "And maybe you should try it too. You'll be surprised what you find."

"You mean soul-crushing sadness?" the blonde asks sarcastically.

"I mean responsibility," the brunette clarifies. "You can't live in your fantasy world forever."

"But I love my fantasy world," Mel whines.

Shree smiles, a legitimate smile. "God, you are so annoying."

The three of them continue to relax on the beach, the sky shading from blue to yellow to pink to red. Brews are passed around, joints, food. Cooper watches the sun sink lower in the sky; already another day has passed. In four billion years, this rock has never stopped spinning. And not even climate change will put a stop to it. The sun will rise again tomorrow, whether there are people around to see it or not.

The sun peaks underneath the horizon, its light obscured by rock. The night blackens; the other beachgoers pack their bags and leave. Only the roommates remain.

A gurgling from deep within Shree's gut. With it comes a wave of nausea, powerful enough that she decides she's had enough beach for now. She gathers her blanket and slips into her flip flops.

"I'll be inside," she says.

Cooper's head cranes back to see her. "You okay?" he asks reassuringly.

"I just don't feel well," she says. "I'm gonna take a nap." Her feet carry her lightly over the sand and back toward the house.

His eyes follow her as she departs. His pupils trace the horizon, passing over bushes, boulders, and finally, to Mel. Her eyes are two shiny pieces of coal, dark and fearsome. But against the twilit background, they glisten invitingly on the empty beach.

"You think she's gonna be okay?" he asks her.

She glances back toward Shree's direction. "She says she feels old. Everyone feels old today."

Her fingers trace shapes in the sand, small concentric circles that look like a mouth. Bored, she looks up at the celestial sphere, the first stars winking into view.

"I don't want it to go away," she says, staring up on this warm night. "But I guess all that's left of childhood is the memory of the way things were."

Cooper looks over to her skeptically. He's never heard her like this before, so ponderous; he didn't think she was capable of such introspection. Maybe she was just hiding from it.

"I think I figured out why adulthood is so hard," she says.

"Why's that."

She moves to face him, nudging herself closer. "Because when you're an adult, death becomes a real thing," she explains. "Every time you hug your friends and family goodbye, you always fear that it'll be the last time." Her eyes glint at him in the moonlight.

He smiles innocently, breaking from her gaze to peer into the starry dome overhead. He considers the splendid points, so brilliant, so unfathomably distant. But he knows that on some cosmic level he's connected to them, a child of the universe, a lonely speck on a lonely speck orbiting a lonely speck.

"I think, sometimes, you get to the point where you just can't figure it out," he says, considering the vastness above. "And all you can do is throw your hands up and say, 'Fuck it, I have no idea. I don't know anything.' And it's in those moments when life is both scary and empowering, moments when you realize you have no idea what you're doing, but you just keep on doing." He looks over at her sensuously beaming eyes. "That's what growing up feels like to me."

She giggles receptively, resuming her wistful glance toward the sky. She truly doesn't know anything, it seems.

"Am I pretty?" she asks the boy.

His head snaps to face her. With moonlight reflecting off her curves, she certainly looks divine.

"You?" he asks drolly, trying to dodge the question. "Yeah, I think so," he admits. Blood flashes up to his head; he's spilled the beans, as if it's anything to hide.

Mel reclines against the towel. "Sometimes I feel like I'm not," she says. "Nobody wanted to get to know me today. Made me feel really unwanted.

Between that, Shree, and mom..."

She stops herself short, not wishing to reveal all her insecurities at once. She looks over to Cooper, who seems entranced by her words. She can almost feel his heart thumping through the sand. In this moment, she sees an opportunity — one final chance for control. She may not be able to have everything she wants, but she can still have this.

"Hey," she whispers, rolling onto her side. "Do you think you'll ever lose your virginity?"

Cooper plays with the sand innocuously. He glances up at the gods in the sky, the sparkling, eternal stars. They'll all still be here after he's gone.

"I sure hope so," he says.

She wriggles closer to him along the sandy beach. "Have you ever even felt a girl's tits before?"

His face flushes, but he hopes she doesn't notice in the evening dusk. He feels his little soldier poke up from his pants, the direction of this conversation piquing his interest. He moves closer to the half-naked girl beside him.

"No, I haven't," he says.

She smiles at him invitingly. Without words, she brings her body right up next to Cooper's shirtless torso. She reaches behind her back and unhooks her bra until it hangs loosely on her shoulders. Cooper watches in slackjawed captivation.

"Go ahead," she breathes.

He blinks and swallows hard, amazed to finally experience the supple roundness of a woman's breasts for the first time since he was weaned. The opportunity has sprung on him as quickly as the erection in his pants, now pulsing with primal urges.

Ever since he hit puberty, he promised himself that he wouldn't die a virgin. That he wouldn't be one of *them*, one of those losers that miss out on the satisfaction of that biological urge to procreate. No, he would fulfill his pledge to his genes, to life itself, to become fully human, to carry it forward. Cooper wouldn't miss out on this act, the incredible sensation of fulfilling the purpose of every cell in his body, every nucleotide — an experience that could be proof enough in the divine. With a beautiful woman in front of him, the time is now. He may never have another chance.

He brings his hand slowly toward her chest, gently lifting the turquoise cups off her body, unwrapping her bosom like a fine present. Beneath her bra are two plump breasts, round as candy apples, crowned by beautiful perky nipples. He runs his warm palms over them with tender desire.

Mel arcs her head back with pleasure as he passes his fingertips over her body. Their juicy lips meet, their naked bodies embrace, and on this starlit beach on a doomed planet, Cooper crosses another milestone on his road to manhood.

Vein of Stars

That Shree has gone several days without making an appearance at work has certainly not been lost on Principal Sanchez. The students need their teacher, and this one is being irresponsible.

The girl enters the room like she's walking into her own funeral, her head dipped low and eyes averted. She sits awkwardly in the chair before the tall, important desk. On the bulletin board is a large calendar with an entire row of days blocked out in red permanent ink, its all-caps shouting like an angry baboon: "TESTING WEEK!!!"

Just from the way the girl is holding herself, Sanchez can glean that something is amiss. It's obvious that she's staring at a strained human soul.

"Why did you call out of work two days in a row?" the principal asks sternly.

The girl's face turns paler than gravy. "I've just needed some time to myself."

"We need you in school," Sanchez replies firmly. "The students need their teacher." The principal is right, and Shree knows it.

"I'm going through a lot right now, Ms. Sanchez," she says. Her hand rubs along her arm anxiously.

The principal has seen cases like this before. New teachers, brought into the punishing conditions of the public school system, completely unprepared for how the world treats them — like garbage. With garbage pay and garbage respect. If Shree is to survive as a teacher, she'll have to toughen up and recognize that teaching isn't a profession of instant rewards. It's a profession of future-building, when the investments won't pay off for ten, twenty years, when despite all the shit you've had to trudge through, someday a student will come back and tell you how important you were to them. It will take a generation before Shree's efforts really pay off, and by then it'll be too late.

"We're all going through a lot, Sharifa," says Sanchez; she's decided the best remedy is a firm pat on the ass. "You're an adult. You're expected to show up for work."

Shree shakes her head emotionally, fighting the tears, though they come anyway. She puts her face in her hands, tactfully hiding the display from her boss.

"I just...I don't think I can handle this," she admits from behind closed palms.

Sanchez, however, is unmoved. The lead-up to testing week is always the most stressful, and it can be the defining moment of a new teacher's career. In every case she's seen, this is when new teachers cross the point of no return. They either make it or they don't. People like Shree always do. Except this time.

After a moment, the girl collects herself. She finally has the courage to say it.

"I'd like to request a leave of absence."

Sanchez eyes her with unflappable detachment, though her internal reaction is secretly one of unmitigated shock. All this time, she had expected Shree to pull through.

"That's probably a good idea," she says gravely.

Shree nods silently, looking down at the floor in shameful disgrace. As a professional, she thinks she should be stronger than this. But she isn't, and she doesn't realize that it's okay. Instead, she bathes in her failure, marinades in it, washes it into her skin like lotion. The stench is horrendous, the smell of loss so foul and dark, poison worse than carbon.

The principal wheels her chair over to the calendar behind her. "Next week is showtime," she reminds her. "I need you here. Your students need you here. If you want a leave now, it will be unpaid. But if you're willing to wait until after testing, I can put you on leave through the rest of the year if you need it, fully paid."

It's a reasonable offer, yet Shree is tortured by the decision. Like most things, the choice is an economic one: "I'll stay for testing."

. . .

April had come in like a lion, the interminable winter dragging on obstinately. But come May, the weather morphed into a lion of another sort, unleashing an infernal heat onto the Northeast. The blistering warmth continues to sear the entire region, New York at its epicenter.

The children sweat in the muggy gym, the air heavier than a wet towel against their bodies. Pencils scribble black circles onto answer sheets, some of them trembling in the fever of the room. Only 26 minutes remain to finish the exam. Don't pass, and their academic futures could implode before their eyes. The pressure is on. Tick, tick.

Shree wipes the moisture from her clammy cheeks. *There's no time for sweating,* she tells herself. The sand continues to fall through the neck of the hourglass, grain after grain plunging into eternity below. Only a few left now, for any of them.

Nervously, she begins tapping the eraser end of her pencil against the top of her left desk, hammering it into the laminate woodgrain like a jackhammer.

"Miss Dunlap?" a student calls out, obliterating the rhythm.

The teacher jumps in her seat, sending her pencil clattering raucously against the floor. She glances at the raised hand. "Yes, Alyssa?"

The girl looks flustered. "Can you please stop that tapping noise? It's really distracting."

Shree blinks twice, then looks down at the pencil on the floor. "Right, sorry honey," she says, grabbing the yellow rod.

Alyssa smiles warmly and returns her attention to her test, driving a tip of graphite into the paper. After making a dark oval on the page, she considers her choice again. Maybe the answer is actually C, not A. She decides to erase the mark.

The clock seems to be growing louder, more insistent. It drums on Shree, pounding her down. The stack of yesterday's answer sheets remains on her desks; she decides to thumb through them briefly, ascertaining where her students stand. Just from the first few slips, she can tell that the faculty will be having another party soon. Not that she'll be there for it.

She pulls out her phone and taps out a text to Pando:

Shree: They don't understand, why don't they understand? I can't make this any clearer. 1:39pm
Pando: thats the question of the ages 1:39pm

This message doesn't hold the reassurance she sought; he never seems to learn that the last thing she wants to hear is doom and gloom.

Enough is enough, she decides. Maybe it's the sweltering heat, but a flash of anger overcomes her. She'll never feel better with a fatalist like Pando in her life. It's time for her to move on.

Back at the apartment, the refugee she's decided to dump has just returned from BKP and is sweating along with her. His perspiration is less about the thermometer and more about the mission he's decided to accept.

Pando is going to lead the bombing.

He looks around detachedly at the mess in the apartment. A mountain of dishes sits in the sink; there's hair and crumbs all over the unswept floor; flies have become permanent residents, as have the cockroaches. Over the course of the year, the place has slowly degenerated into a sty. This is what happens when children rule the house.

He can't help but feel it's his fault. The stubborn cold war of waiting to see who will clean first has reached an impasse.

Just do it already, he tells himself as he bears on the crowded sink. Most of the plates are Mel's, but it doesn't matter. Now that revolution is days away, he may never have a chance to do this again. He's throwing in the towel on this fight, not to quit, but to do the right thing. The water starts; the war ends. With a squirt of soap, he begins to wash the soiled plates.

As he laves the dirty saucers, he realizes this would be much more fun if he was high. He pauses to take a rip from his pipe, climbing into an altered state of awareness. He brings the ashes to the trash.

A gruesome smell fills the room as he opens the wastebin. "Wow," he exclaims, overwhelmed by the stench. "Alright, *that* smells like diarrhea."

He looks into the trash bag, full to the top with decomposing waste. He had urged his roommates not to change the bag until it's full, to cut down on plastic waste; maybe that wasn't such a good idea. He replaces the bag and returns to his work.

Shree's gonna be so pleased when she gets home, he reckons. Maybe, if he cleans the whole apartment, it will brighten her mood. It might even make her day. With renewed motivation, he gathers some supplies and hunkers down for the deep clean.

By the time Mel returns from work, Pando is still scrubbing the kitchen. She watches him clinically, observing him like a monkey in a cage. That he's cleaning almost insults her, like he's sticking it to her by cleaning *everything* so thoroughly like this. But recognizing how irrational that idea is, she disarms, softening her stance.

"Aww, thank you," she says to him, alerting him to her presence.

He glances up from the floor. "Hey," he says.

"You're cleaning all this?"

He nods with a negligible smile.

"You're the best," she says appreciatively.

He chuckles. "Nah, not really. I just wanted to do some good for once." He resumes scrubbing.

Mel considers her ideological opposite on the floor. They may have differing opinions about the things that really matter, but they have more in common than she likes to admit. They're both very proud, very self-assured. And watching Pando step up to the plate before she did, well, that's pretty humbling. In this moment, she feels rather inferior.

"You want me to get the bathroom?" she offers, trying to save face.

"Don't even worry about it, Mel. I'll take care of it all." He stands to look her in the eye. "I cleared my whole schedule for this. Which basically means I went from doing nothing to doing something." He smiles at her warmly.

"Okay," she says with a laugh before retiring to her room. Pando, high as a weather balloon, pulls out his dumbphone and starts playing music. This isn't so bad after all; he just had to get started.

He looks around at the disgusting floor. Bottle caps, wrappers, pieces of paper. So much to clean. He isn't daunted, considering each article of filth as just another challenge. He'll erase them all, and leave a gleaming white behind. This will be his masterpiece in porcelain and linoleum, a brilliant work of art.

Back at school, the day is drawing to a close. Testing is finally over, and Shree has reached the end of her tenure. She hung on for this long. Just thirty minutes left till freedom.

The students are as exhausted as she is. She's supposed to be using this time to teach them a lesson on the Fertile Crescent and the rise of civilization, but she can see by their faces that they aren't having it. Their minds have turned to cheese, and so has hers.

"I'm gonna go off topic for a bit," she says mid-lesson. "I want to talk to you all about something."

She gathers her students into a circle on the gym floor, each of them sitting crisscross-applesauce as they look at her with anticipation.

"The world is changing, boys and girls," she begins. "There are some things happening that you should know about."

She gives it to them straight. By the time they're her age, the condition of the planet will be noticeably worse. By the time they're old and handicapped, they may not even recognize the place at all.

"So what do we do about it?" an innocent voice speaks up.

Shree looks sullen. "I really don't know, to be honest. I really don't know." She considers everything that's happened, all the pain she's had to suffer. And then, without inhibition, she begins sobbing quietly in front of everyone.

The kids are speechless. They don't often see grownups cry, let alone their teacher.

"But I do know," she concludes, regaining her composure, "that I'm proud of you guys. And I want each of you to go home tonight and hug your parents and tell them you love them. And maybe you can talk to them about it. Maybe they know what to do."

Alyssa, the sharpest girl in class, raises her hand. "What if we all recycle?" she asks naively.

"That would be a good start," Shree encourages.

"I bet I can recycle more than you," taunts one boy. Alyssa replies by sticking out her tongue.

"My mummy buys vegetables from the local garden in the old abandoned lot," says another girl.

"Yeah, and my dad always tells me to turn off the lights when I leave a room," says a typically silent boy.

"There's lots we can do," another chimes in.

Shree smiles. Maybe there's hope for the future after all.

"There's something else I wanted to talk to you about," she says, folding her hands. "I'm going to be away for a few weeks while a substitute comes in to take my place. Hopefully I'll be back soon, but she's going to be taking care of you guys for the rest of the month."

They stare at her like she's just announced the end of the world. Nobody wants her to leave. But before she can elaborate further, the end of the day announcements blare over the intercom.

"Congratulations to all sixth and eighth grade students who have finished their state exams," says Principal Sanchez. "All bus students are dismissed."

Nobody moves. The students gawk at their teacher, waiting for something they know will never come.

"Anyway, that's it, class," she says, breaking the uncomfortable silence. "We made it to the end. You can go home now."

The students look at one another. Slowly they get up and grab their bags, filing out of the room in a line. One by one they say goodbye, wishing Shree a happy vacation. They don't know it's not a vacation.

Alyssa approaches Shree before she departs, the last student in the room. "I'm going to plant a tree in the park tonight," the girl beams, "and I'm gonna make the world better."

Shree smiles broadly and gives her a meaningful hug. "I know you are," she says.

Satisfied, Alyssa turns for the door, smiling. "Have a great weekend Miss Dunlap!" she calls as she skips out.

It won't be enough, Shree thinks. *But it's better than nothing.*

She has a hunch she'll never see her students again.

After such a draining, punishing week at work, Shree should relish going home for the night, but she's actually dreading the thought of it. The time has come for her to purge all the negativity from her life, starting with the worst offender: Pando.

They've grown considerably more distant in recent weeks, the refugee having completely devoted himself to the failed drone mission. But despite whatever physical benefits they may gain from each other's brief company, Shree knows it's outweighed by his incessant pessimism, dragging her down like a curse. After mulling it over carefully, she's finally decided to expel him from the apartment.

The commute home certainly doesn't put her in a better mood. She misses the train by mere moments because a pair of slow-moving tourists blocks her way, forcing her to wait on a platform that smells like spoiled milk. She watches someone discard a crumpled brown bag onto the track like an idiot, as if paper isn't readily combustible on sparking train rails. Finally she watches an asshole kiss his girlfriend goodbye, only to immediately pull out his phone and start swiping on a dating app. It seems like the entire body of humanity is populated by stupid lying fuckwads that don't give a shit about anything that actually matters. She doesn't want anything to do with any of them.

By the time she returns home, her feelings are in total control. Her dissatisfaction with everything has crested to such an extreme that she has to fight for composure as she ascends to 4G. When Pando asks her how she's doing when she walks in, she brushes him off and climbs to her room alone.

"Okay then…" Pando says dejectedly. He remains on the couch, giving her a wide berth in case she needs some alone time. She didn't even notice the clean apartment.

"Pando!" she shouts from her room. "Can you come up here please?"

He duly obeys, mounting the rope ladder up to the loft. He enters her room respectfully.

"Hey," he says quietly, his arms held open for an embrace. She sits on the bed instead.

"We need to talk."

His arms collapse to his side, his spirit deflating like a popped tire. He takes a seat.

"I don't want you here anymore," she begins.

Pando's focus dissipates into the stale air. Something has loosed itself in his mind, like a guywire has snapped, toppling his brain like the funnel of a sinking liner. As he listens to her explain why she doesn't think he's a positive influence on her life, his mind melts away, and a paralyzing numbness washes over him. Maybe his approach this whole time has been all wrong. Maybe he's been part of the problem after all. Maybe people don't want to hear the blunt truth. Maybe they want to live in their lies.

"By this time tomorrow, I'd really like you to be packed and ready to go," Shree concludes.

The man's soul collapses into a pile of ash. Ever since Powell, the hardening of his heart has managed to keep everyone out — everyone except her, who snuck in unnoticed. And now that he's lost her, sorrow is all he knows. Sorrow the likes of which he hasn't felt since that day in November, when the sea claimed his brother, his family, his life.

Now he truly has nothing. No home. No future. Nothing but the clothes on his back. And that just isn't enough.

. . .

Cooper can overhear Shree's sobs through the wafer-thin walls. It isn't the first time he's heard her cry like this; usually, however, she's had Pando to comfort her, if comfort is something Pando even knows how to provide. He decides to fill in.

"Hey Shree?" he says, knocking on the closed door.

She opens the door with a sniffle, eyes bleary and cheeks sodden. She motions him inside and closes the door.

"Is there anything I can do?" he asks, rubbing his arm.

She rubs her brow forcefully, confusion on her face. "I don't even know," she says. "I think I'm having a breakdown."

He takes a seat beside her on the bed. Instead of talking, he just listens.

"I always thought the best thing I could do for the world was give back to it," she says. "But now I—"

A buzzing chirp from her phone breaks her line of thought. She grabs it instinctively, swift thumbs tapping the screen. She returns it to the mattress.

"What was I saying?" she sniffs.

"Giving back to the world," says Cooper.

"Yeah, I just thought that the whole point of all this was to leave something behind, something to be remembered by, right?" She looks at him for an answer. "Isn't that the point? To leave the world a little better?"

He doesn't say anything. Just being there is enough; he doesn't need to answer.

Another cheerful trill vibrates from her phone. She pulls it up to her face, reading it with beady eyes. She writes again.

"But now," she goes on, "it seems like the only contribution I've made..." She whimpers, fighting tears. "I thought that good people have good things waiting for them. But now I know that all of this..." she motions around her, "is coming to an end."

The bright tone from her phone sounds again. After another moment, she returns to physical reality.

"You know what I mean?" she asks. "I thought being a teacher was gonna be a great, meaningful life. But there's no meaning in this. There's no future. It's an illusion. And whatever reasons I gave myself for living..."

Her face darkens, her eyes turning down. She shuts them forcefully, the tears evacuating from her eyes.

"...they're just lies."

Cooper feels his heart sagging, the weight of her words almost too much to bear. If Shree can be dragged down like this, plummeting off an emotional cliff, then no one is safe. It can happen to anyone.

He opens his mouth, finally ready to respond.

The phone sounds again, almost louder this time. Every notification chafes Cooper's ears, the stupid digital chime contrasting with the existential murk shadowing the room. Whoever the hell is on the other side of that phone better be fucking important.

"Who are you texting?" he asks peevedly.

She taps the screen rapidly. "My mom. She's worried about me."

He waits until she's done. "Listen," he says, placing his bony hand on his friend's shoulder. "I know how you feel. I've been struggling with the same questions all year."

She considers him through tears. "So what's the answer?" she asks, as if she's posing the question to God Himself.

He shakes his tiny head. "Just because this won't last doesn't mean it has no meaning," he offers. "It has meaning for us here, now. Of course it won't last. But for a fleeting moment in time, we lived in a country that at least dared to be better. That at least tried to do the right thing."

He rubs her shoulder affectionately. A shaky smile begins to crawl across her face. She's made up her mind.

"Whether we get it right or not maybe isn't as important as the fact that there were those of us who tried our best," he says. "And we just have to keep trying our best."

The smile blooms onto her face. That he failed to convince her isn't important; at least he tried his best.

"You're right, Coop," she says, wiping water from her face. "At least we tried."

They hear the apartment door creak open and closed. Pando is home.

"Well thank you," she says, putting her warm hand on Cooper's. "I think you should help Pando with his things." She brings his hand back to his lap.

Downstairs, Pando is gathering his few remaining things into a bag. After 24 years on this Earth, his personal belongings amount to a toothbrush, a couple sets of clothes, a cellphone, a wallet, a computer, a pipe, and a small bag of drugs. Not much to show for a life.

"Hey man," says Cooper from the loft. He makes a start down the ladder.

"Hey," says Pando, now completely packed; it didn't take long. "You ready for tomorrow?"

Cooper walks up to his colleague dearly departing. "Of course, buddy."

Pando nods somberly. "Don't forget the speakers." With backpack slung over his shoulder, he leaves the apartment one last time.

"Wait — Pando." Cooper follows him out the door.

"What, you're gonna say goodbye?" asks Pando. "You know this ain't goodbye."

"I know," says Cooper, closing the door behind him. "I just...I just wanna know what happened. With the breakup."

The homeless man shrugs. "She said we weren't good together, that my hatred of people was driving her insane."

"She's right ya know," Cooper lambasts. "You do hate people, and you can't."

The bag falls from Pando's shoulder, thudding onto the floor. He looks at Cooper desperately. "Why not?" he asks.

"Because hating people is too easy."

"Too *easy*?" Pando fulminates. "You're right, it is too easy. With all their deceit and denial, living in a fantasy world while destroying the real one. They've ruined everything that made this a place worth living in." He grabs his bag once more and makes for the stairs.

Cooper watches him down the steps. "Where are you going?" he calls out.

Pando halts his descent and stares up at Cooper, his eyes like metal drills. "Out." In other words, he'll be sleeping in a park tonight.

"I'll see you in the morning," he adds. And then, like the soldier he is, Pando resumes his march, down the stairs and out the door.

The Drying of the Lawns

The train bumps along rickety tracks, a silver steed propelling the young men through the ancient tunnels. This is it. Things are about to happen. He and Pando are en route to BKP, but they've decided to spice up their commute with a little song and dance before they walk onto the battlefield.

As soon as they entered the train, Cooper pulled out a pair of wireless speakers from his backpack and placed them on the floor between two opposite doors. With Pando now ready to perform, Cooper pushes a button from his phone. An ill track begins to play, drawing curious eyes in their direction.

Pando starts moving to the beat, then, when the moment is right, jumps in:

> Listen up now, y'all, just give me your eyes
> Forget all the phonies in their suits and their ties
> The ocean is coming to swallow us whole
> While the leaders do nothing, our future they stole
>
> They said we'd thank them when we got older
> But now we must arm, a generation of soldiers
> Our elders just use us as their carbon fodder

While life on Earth bleeds and our planet gets hotter

As the refugee piles on the passion, Cooper realizes he's witnessing the advent of what can only be described as psych-hop, a transposition of the swirling sounds and synthesizers of the psychedelic sixties into today's contemporary tastes and production. The birth of a new genre, a message in its soul.

The audience has taken notice, rubbernecking around each other to catch a glimpse of the young preacher at full birr. He raps faster, packing on the emotion, belting words into the air like shotgun blasts.

They deny the truth, mere puppets of felt
But we totally uncool with this hand we been dealt
Our teeth we cut, our swords we smelt
With the passion to save the glaciers they melt

They nag us senseless, shoot us in the street
While the machine drones on and the war drums beat
But we ain't asleep, we gonna stop the heat
If they wait too long, they'll lose their seat

This train is moving, we delay no more
We kids have grown smarter than they'll ever know
We've come to collect the dues they owe
And turn our backs on the future they've sown

So get off your damn phones
And pay attention for a bit
The world has turned to stone
And it's time to give a shit

He finishes with a crescendo, his voice seized by some unseen angel channeling a divine spirit. He looks about in the heavily-peopled train car, all eyes beholding him. Phones are up, their cameras aimed at his face. At his conclusion, the crowd detonates into applause.

Cooper claps too, thoroughly impressed. "Not bad," he winks.

Pando smiles at his partner with prideful confidence. "I had to make sure I did that at least once in my life."

They disembark from the train and climb the stairs to the street. Now that it has service, Cooper pulls out his phone to scan his notifications: an air quality alert, a few social media Likes, and an article highlighting the international climate conference set to commence later this week. "Current pledges by industrialized nations not enough to keep warming to two degrees; 4C now 'all but certain'," it reads. In other words, humanity is about to fail.

Yet the sultry Brooklyn air seems to spark with tension. Parked on the street is a car with a BKP bumper sticker: "Do we really want to live *here?*" it asks, lain over a burning globe. It isn't the first Cooper's seen; everywhere he goes nowadays, he sees the BKP logo — stamped onto lightposts like prison tattoos, chalked into dusty pavement. All around him, a movement is beginning to coalesce.

He can feel the needle finally starting to budge, can smell it in the air. It smells...like soot. Like burning carbon. Ash somewhere, maybe from a cigarette quashed against the cement, or someone smoking a bowl on their patio.

It's more than just the mood. It's the way people talk, the way they move. There's a general disease in every body Cooper can see. They're stiff and defensive, like brutal hands have seized them by the shoulders. These humans have fear in their eyes. People speak in whispers, afraid their voice might be overheard by someone with an opinion. Public demonstrations happen daily, though outnumbered two-to-one by the mass murders. The violence feels omnipresent, like CO_2 in the sky. Maybe the heat is getting into their heads, colonizing them like mold. Or maybe they've finally started paying attention.

Despite the pervasive fear, however, there is also the hope — that the days, months, and years ahead are going to be ones for the history books. That the world is entering a new chapter. Teach-ins have been planned, marches scheduled. Heroes are emerging, beacons of foresight shining wisdom onto the masses like golden drops of holy water. Cooper sees them more and more, everyday people taking an interest in the climate. Young people everywhere are championing their rights, bringing the fight to every corner of the country, from Florida to Alaska. These are the Malcolm Xs, the MLKs, the JFKs in their prime. These are the sixties.

By the time he's seated in the meeting room, the impending climate conference seems an abstract irrelevance, for before him on the giant screen is an overhead view of the doomsday geoengineering device — or lack thereof.

"As you can see," Iri illustrates, "the geobomb is gone."

The photograph is one of the dozens from Pando's recon mission. There are no aerosol canisters to be seen.

"Do you have any leads?" Cooper asks.

Iri shakes her head no. "We obviously can't afford to wait for it to resurface," she says. "We have to reveal its existence now. But we can only do it at a moment when the public is solidly against the Cokeheads and the carbon they burn. And the only way to do that is with a bomb of our own."

A gagging sensation engulfs Cooper's throat like an eggplant in his windpipe. He looks over to Pando, who sits beside him in pensive attention.

"The target is the Columbia divestment protest," she explains, flipping to the next slide. A familiar map of campus diagrams a stage set up on the university lawn, before which hundreds, potentially thousands of people will be demonstrating. "A small IED under the stage will claim a life or two, the public will go berserk, and we'll frame the Cokeheads for the carnage. By the time we tell the world what they're planning, revolution will have already begun."

"You're gonna bomb *Columbia*?" Cooper asks, flabbergasted.

Iri nods stoically. "And Pando, you have volunteered to be the bomber, correct?"

The refugee bows his head. "I'm the one for this mission," he says. "I've always been the one."

"And Cooper," she says, locking eyes with him, "you're gonna be his backup."

The fledgling adult grimaces at her. "Me?" he asks. "No, I can't." His worried mind drifts to thoughts of Natasha. All her hard work is about to end in disaster.

Iri frowns angrily. "Cooper, you pledged your support to us, that you would help us bring down Coke," she reminds him.

"But I can't," he pleads. "I can't put people in harm's way. Not like this."

"They're already in harm's way!" she shouts.

She grabs him by the back of the head and wrenches his face toward Earth 2100 on the wall. Its blood-soaked globe draws his thoughts, his fears combusting into a waking nightmare, sucking him into dystopia for all eternity. A death sentence. Terminal cancer. The banner before him is the embodiment of failure, of evil's triumph. It is a world of climate catastrophe. It's a world he'll die in.

"Look at it, Cooper," she snarls, gripping his head like a vice. "Do you really want *that* to happen? *Can* you? Could you even live with yourself?"

Emotion shudders through her body and into his, the rippling passion of a limitless universal respect for life, so radiant and compelling and utterly divine that she has no choice but to believe in it. A force so powerful, so thoroughly irresistible, that she is willing to risk it all for the defense of her principles, the principles of life.

Cooper remembers the figures, the facts. The irrefutable proof. The toxic future. Despair again takes hold.

"No," he says, because what else was there to say?

She releases him. "Didn't think so."

He presses his fingers against his forehead, a futile effort to stave off a headache. He knew there would be bloodshed, but his own classmates?

No, this isn't right. But he swore an oath to protect the future, a promise that, for the sake of his soul, he intends to keep.

"What do you need me to do," he relents.

"Make sure Pando completes his mission," she insists. She zooms in on the diagram, eliminating from the frame all but the stage and a few rows of the crowd. "The bomb will be remotely activated, but it can only receive a detonation signal inside a fifteen-foot radius, which is within the blast zone. Pando will have to stand at the very front of the crowd if he's going to succeed."

Cooper turns to his roommate distressingly. Pando's face shows no emotion, stone as always.

"Sounds a lot like a suicide mission," Cooper laments.

"We live in suicidal times, Cooper," Pando responds. "And if the Anthropocene is an age of suicide, then let mine be an act of defiance in the march towards that age, that my sacrifice may avert the disaster that you and I know must not be allowed to unfold."

Iri turns to her hand-selected soldier. "We've furnished a manifesto that will upload to your online profile half an hour after the attack," she tells Pando. "It will say you were following marching orders on behalf of the Coalition."

Cooper considers the plan. Risky, to be sure, but what isn't these days? Taking a long shower is risky. Eating a thick steak is risky. Driving an SUV to work is risky. Killing innocent civilians and blaming the Cokeheads? Risky isn't even the word; it's revolutionary.

The slide changes again, exhibiting a diagram of a crude IED. Iri describes in detail the design of the tiny bomb, how when Pando pushes his handheld trigger, a spring-loaded switch will sever a wire in the explosive, leading to detonation. Cooper burns the image of this schematic into his memory; knowing how bombs work could come in handy as a wanted vigilante.

The divestment rally is mere days away. Natasha will probably be on that very stage, a mine lurking beneath her feet. And despite the gnawing protest of his conscience, Cooper allows himself to accept his duties in this mission. Arguing with Iri is as useless as arguing with the deniers; BKP will follow through with their plan no matter what he does. His lack of agency has led him to this moment, sleepwalking into a role he knows could be the worst thing he'll ever do. Maybe even the last.

The meeting adjourns with solemn urgency after an hour of maps and strategy, covering things like where to stand, how to act, and how to respond if they encounter a Cokehead or two. Neither agent displays any hint of protest. Pando in particular has accepted his role in history as a martyr, to make the ultimate sacrifice for a cause greater than self. They step together into the blinding light of day.

"So this is how we save the world," Cooper mutters, kicking a red bottlecap down the sidewalk. "And I guess this is it for you."

The refugee displays no feeling, as if he hasn't yet processed the full consequences of their plan. Non-existence must be hard to think about.

"Are you comfortable with this?" the youth asks his guru, searching for reassurance.

"Of course I am," he replies. "Iri didn't want it to be, but after Shree kicked me out, I requested a one-way trip. It's the only way I could go through with this."

"You mean you *asked* for a suicide mission?"

Pando nods modestly. "People are gonna die, Coop, whether it's from you or me or the Cokeheads or climate change." He finds the red cap and kicks it further down the street. "If I have to be the one to set things in motion, I better be willing to pay the ultimate price. And I am."

"People have *already* died," Cooper adds. "But this feels wrong in every way."

Pewter eyes flash in his direction. "You wanna do something, don't you?" he presses. "Well this is it. This is all that's left to do. This is all they've left us with."

Cooper envisions the carnage, the rivers of young blood. He sees Natasha, a beautiful soul, cut down in the prime of life. And what he can't see, despite all efforts to convince himself, is a better future on the other side of it. All he can see is a black pit, the world shriveling like a rotting banana.

"I know some of the people that will be there," he says. "We're probably going to kill them."

Pando looks over at him with surprise. "Ya know, even after all these months, I still don't understand you, Cooper," he condemns. "I don't understand anyone. It's like you aren't even people anymore. You find it so easy to weep for the loss of one life, when a billion more pass by unnoticed."

"And I don't understand you," yells Cooper, "for feeling nothing for that one life."

Pando pushes his hand into Cooper's shoulder, swiveling the boy's torso to face him. "How can you even say that?" he admonishes, a clenched fist waiting by his side. "Do you even know what Powell was like?"

The Floridian stands in place like a statue, awaiting the inevitable diatribe, the white flecks of spittle falling on his face like bird poop.

"I have to tell you something, Cooper." Pando's fist relaxes, and the anger in his eyes subsides. "The typhoon didn't kill my brother."

Cooper cocks his head curiously. "What do you mean?" Has he been lying all this time?

He stares into Cooper's eyes. "I did."

The words freeze Cooper's heart. This isn't a refugee before him; this is a murderer. Then again, so is everyone.

"He was pinned under a log," Pando says, recalling the painful memory of his brother's leg skewered by a twisted trunk of wood. "I could've helped him. But there were a pair of kids in the water, drowning, being swept out to sea. They weren't gonna make it either, not on their own. I had to make a choice." He hangs his head low, his eyes shadowing over as he remembers the murderous scene from half a world away.

"So you saved the kids instead," Cooper says, understanding.

"I figured two lives were better than one," he justifies, his voice modulating into somber notes. "My mother didn't seem to agree, when I told her what had happened. That I could abandon my own brother like that. She said she'd never talk to me again. Hence why I'm here, and not there."

Finally Cooper begins to understand the vastness of Pando's desolation. Unable to save his family, and unable to save the future, the poor man's loneliness knows no bounds.

"Which is why I'm okay with this," he continues. "There's already blood on my hands. If anyone else has to die, it might as well be me."

The death wish reveals itself at last.

They continue to walk, passing by an advertisement for bouquets. "It's Mother's Day," Cooper says meekly, noting the date on the sign. His mind recollects the storm that took his own mother all those years ago. Pando's situation really isn't that different.

"Just promise you'll be there for me," says Pando.

Cooper looks at the brassy sky, the sparkling monoliths of New York piercing it like brilliant swords. "I'll be there," he swears.

A taut silence establishes itself between them as Cooper sits with this commitment. He *will* be there for him. He'll be there till the end.

"I miss my mother," he says after a lengthy pause. "And I bet you do too."

Pando looks at the kid, a brother of a different kind: not merely biological, 99.999% similar down to the base pairs, but ideological, a kinship that transcends the boundaries of time and space.

"We still have our mother," he tells Cooper. "And we can still save her."

"Yeah, by killing people."

Pando looks around in subterfuge, making sure no one is within earshot. He pulls Cooper close by the arm. "Listen, man, history is all about timing. What you do ultimately isn't as important as *when* you do it. If avoiding catastrophe for the next twenty generations of human life comes at the price of two or three lives today, especially my own? Hell, I'd call that a bargain."

Cooper tries to rationalize this, but it fails to add up; logic doesn't seem to apply to violence. He shirks off Pando's hand restlessly.

"So you've decided to kill yourself to pay off your debt to the universe," he accuses.

"Better than suffer through Hell with the rest of you," Pando counters. "In a suicidal world, only the suicidal are sane."

"I thought you weren't suicidal."

Pando shrugs blankly. "There's a difference between suicide and sacrifice. I'm not trying to escape the world. I'm trying to save it."

Lucifer's Eyes

A large curved table dominates the conference room at Mel's workplace. Through the floor-to-ceiling window, she sees the crystal citadels of New York surging into the air. She taps her fingers against the black glass of the table in an uneven rhythm, a mood of agitation saturating the atmosphere.

"Profits continue to slide," says the boss figure at the front. "The culprit, as you probably know, is climate change."

Mel pricks her ears at the phrase. In eight months at this job, climate change has been the elephant in the room that no one wanted to discuss. But with the ongoing drought disrupting America's breadbasket, its effects can no longer be ignored.

"The never-ending winter froze much of the crops in the Southeast," the woman continues, motioning to a map of the continental US. "Our harvest is projected to be down 22% this year."

The young secretary peers at the graphs on the screen that have garnered alarm from the whole room. She decides to raise her hand.

"Yes, Melanie," says the boss.

"Do you think this could postpone the opening of our Brooklyn office?" asks Mel.

The manager adjusts her glasses. "I'll get to that in a few."

And she does. Through graphic descriptions of the condition of the nation's agriculture sector, the boss explains the ravaging woe that extreme weather is having on their industry. Their plans for expansion will have to be put on hold as the relevant people assess damage to their clients' farmlands. Instead of upsizing, the company now appears to be on the verge of contraction.

At its conclusion, Mel storms out of the meeting, flailing in the midst of a quarter-life crisis. The feeling transcends unhappiness; it is something more akin to existential panic. Her expectation of an imminent promotion has been dashed by cold reality. She realizes, for the first time, that this global warming nonsense might not be a casual irrelevance after all. It may, in fact, be the single biggest threat to her career, and with it her livelihood.

Returning home that night, she fires up the bong that Marquis left in her room during his last visit. She packs two bowls: one for her, and one for her fallen friend. As she smokes, her thoughts coalesce into a vision of his death, every detail unfolding in slow motion — his car colliding with unyielding wood, aluminum folding and deforming, a fault line splitting open a pane of curved glass, fracturing into a million different directions. His skull is slowly crushed under the force of the impact, popping like a blister. Tomato juice flies everywhere.

Her thoughts flash simultaneously to the data points her boss presented in the meeting — the perilously climbing graphs; the red on the maps; the gradual, monthly minutia of planetary collapse, a film which, when played at geological speed, happens in the blink of an eye.

Incandescent rage flares up in the hollow of her chest, shooting up through her mouth into a penetrating scream that bayonets the walls. Her frustration is pure and absolute. Once a devout agnostic on the issue, she can feel herself bending to the interminable will of the weather. She now seems to grasp the full meaning of where the future is heading, with climate change looming unmistakably on the horizon like a dark, shapeless cloud. Not only is it barring her from a promotion, but it may put her out of work altogether.

"Fuck you, world," she blasts. "You don't give a shit about me."

"Fuck you, too," the world replies. "You don't give a shit about me either."

She's always been a realist. But in this new situation, she begins to understand that the only practical thing to do is act, to challenge the ultimate challenger. If global warming affects the bottom line, then it's a problem; in the end, it's a simple matter of economics. Climate change must be stopped, no matter what.

. . .

At 21 years of age, Cooper's been around long enough to know that his opinion of anything can change. But he's also strongly aware that he has moved beyond the critical period of radicalization. His brain is no longer a fertile incubator on which to graft ideas of any form, but a measured article that filters out all the scrim and screens everything for bullshit. The result: nothing passes the test. Nothing can breach the yawning emotional void between the boy and his world. Nothing elicits response, because his care meter has long since shattered from overpressure. And as such, any idea fed to him quickly ricochets off that unbreakable wall.

The curtain of adulthood has finally been drawn, and unveiled beneath is a steaming shit known as the world today, along with the ghostly premonition of the world tomorrow. As long as the gears continue to turn, and until the machine shows even a moment of weakness, there seems to be no point in opposing it. It would be futile; wasted energy.

His angry young man days have come to an end before they even began. You can only be angry about the world for so long. Anger instills action, but he has no passion to act, no spark. Civilization has already failed. Pain gives way to sheer disappointment, a disappointment which allows him to live in this world without being satisfied by it. Maybe this is the true lesson of growing up — being able to accept the things that lay outside of his control.

Which is why, upon reading the news article on his screen, he finds himself feeling nothing. Scientists have released a sobering report about the state of Earth's ice sheets, concluding that the complete destabilization of West Antarctica is already underway, and likely to continue unabated. The sea will inundate whole cities, from Miami to Manhattan, no matter how quickly the economy decarbonizes. It looks like the climate conference is too little too late; the planet has slipped past a crucial tipping point, and now seems to be awaiting its ultimate fate.

The timing couldn't be worse, if there can even be such a thing as a well-timed Apocalypse. Graduation is slated for next weekend; the end days are fast approaching. In a few weeks, Cooper will enter the "real world," a world peopled by phantoms and lies, all of which are about to come into stark relief. All the bullshit he's been fed through the course of his life is about to buckle like a house of cards. So many lies. That he's special. That he has a bright future. That college prepared him for something. That he'll retire someday. That his best is good enough. That he has something to contribute to the world. That there's still hope.

He visits another news site, prepared to read more about the inevitable Antarctic collapse. Instead, the headlines disappoint. The Middle East descends into chaos. The global economy underperforms. Yet nowhere to be found is any mention of the climate. Cooper's educated eyes scan the page thoroughly. By omitting Antarctica from the headlines, the media proudly displays another middle finger to the planet. Really, to everyone. Just another missed opportunity.

It doesn't matter anyway. In 24 hours, he and Pando will carry out their mission. They'll have crossed their own point of no return, entering a whole new hell of their own. In all honesty, it'll be a miracle if Cooper survives to graduation. Pando, meanwhile, will be dead within a day.

The boy closes the maw of his computer, replacing it on the floor beside his mattress. From his table he grabs *Futility*, opening to the cloth bookmark within. He gets lost in the pages.

An hour passes. Cooper, absorbed in the novella, allows himself to forget about larger concerns. Then a caustic vibration shudders him back to reality, his phone buzzing into the hardwood floor like an angry hornet.

Pando: bkp at 1 for training don't be late 11:43am

Cooper checks the time tag; he'll have to start getting ready now if he's to make it in time. His limp body protests; he doesn't want to go. But he has a duty to uphold, not just to the world, but to his friend. He tosses the book back onto the table, watching it slide precipitously close to the edge before just holding on. He rises from the bed and starts to get dressed.

Phone? Check. Wallet? Check. Keys? Check. He heads out the door, only to hear the sound of something drop behind him as he exits.

He turns round briskly. *Futility* is sitting on the floor.

"These are for you," says Iri, handing two loaded pistols to the agents. "I doubt you'll have to use them, but it's good to be safe."

Cooper accepts the weapon, his face blanched. The semiautomatic pulls his palm down with unexpected heft; he's never held a gun before. It hardly makes him feel any safer knowing he can take a life just by flexing a finger.

"What's the latest on the Cokeheads?" asks Pando.

"Nothing good," Iri frowns. "One of our agents was ambushed last night, which implies they have some way to track our movements. As always, keep the digital communication to a minimum."

Cooper gulps loudly as he thinks about the slate in his pants pocket. If the Cokeheads needed a tracking system, hacking phone's GPS would be the obvious method.

"Do they know what we're planning?" Cooper asks.

Iri shrugs uncertainly. "No way to know at this point. Hence the extra firepower."

Cooper looks back at the pistol in his hand.

"I want you guys to go to the shooting range," she instructs, handing them an address on a scrap of paper. "Be back here by 4 o'clock."

The agents make their way to a disheveled garage in East New York, where Pando shows his young friend the basics of firing a gun. They practice aiming at a black torso-shaped outline on a sheet of cardboard at the far wall, plosive shots passing effortlessly through the fiberboard and digging into the partition behind. How many people BKP has trained in this small warehouse is beyond their knowledge.

This all feels too real for Cooper. Now that he's armed, the danger is palpable, physical, *real.*

"Not bad, buddy," says Pando, regarding the bullet holes in the cardboard torso. Cooper says nothing. The quietude speaks for him.

A pulse from his pocket alerts him of a notification.

Natasha: Are you coming tomorrow??? I wanna see you there! 😊 2:18pm

A quickening of his heartbeat signals there is life in him yet. His throat constricts, his breathing labors. Natasha. Beautiful, beautiful Natasha. She wants to see him. He means something to her after all.

He can feel the color drain from his face, the nausea welling up from within. In all likelihood, the girl on the other side of the text message will be injured or killed tomorrow. Can he really allow that to happen?

His back straightens unexpectedly, the first sign of a spine he's felt in months. After an entire winter of feeling like a festering corpse, vivaciousness springs into Cooper's mature body, growing into every cell like shoots of bamboo. He *is* still alive after all. And he still has a reason to live.

The nausea dissipates, replaced with a dynamism that has lurked for too long. Blood inundates his reddening cheeks. He feels himself bursting from a cocoon, passion spouting through the cracks like geysers. Manhood reclaims his body, entrenching itself there like gum in a sneaker. His muscles snap to attention, his eyes blare. The rigor mortis sloughs off his frame, replaced with the electric vitality of life.

Pando is ready to walk to his grave. And with Shree in emotional freefall, and Mel spiritually dead inside, Cooper finds himself the only one who still cares whether or not he draws breath, still cares that he has a beating heart. And as long as that organ in his chest is functioning, he can still do something, still be something.

Finally, he knows what he has to do. What an adult would do. What a human would do.

"Pando," he says.

The refugee looks over to him, standing like a man, brimming with the vigor of manhood.

"We don't have to do this."

"What do you mean?" Pando asks.

Cooper points to the gun in his hands. "This," he says. "It doesn't have to be this way."

Pando closes his eyes in irritation. "No, Cooper, it doesn't have to be this way. But it *is* this way. So live with it. Man up and live with it."

"No, I'm not gonna just *live* with it," Cooper snorts. "I'm not gonna kill people just because a bunch of radicals told me to."

"You're not gonna kill anyone," says Pando. "You're just gonna help me do the dirty work of the human race. That hardly seems radical to me."

"So what? So we can save a future that's already lost?" Cooper's voice is more forceful now, more directed.

"So we can fight!" Pando enthuses. "So you and Iri and BKP can take the war to those Cokeheads and stop them dead in their tracks. It's a war, Coop, and the bad guys are winning. Don't you ever forget that we're on the side of life."

"We can't win a war like this," Cooper scowls. "Do you realize how many people we're about to hurt? At best we'll end up in prison, and at worst we'll both be dead."

"At worst, we'll *all* be dead, so don't even start with me," Pando says.

"We can't save the human race by throwing away what it means to be human," Cooper pleads.

"Then I guess humanity is already lost!" he fires back. "Because only a race of nonhumans would choose to kill everything and everyone."

"I'm not doing it."

Pando realizes he's witnessing the early stages of Cooper's defection. His eyebrows jump, hysteria bleating through his senses. The boy can't back out now. He can't.

"So what are you gonna do?" Pando taunts. "You're just gonna serve coffee for the rest of your life? You're gonna serve coffee while the world burns?"

Cooper's mouth closes, his face despondent. His hands are shaking, the anxiety bubbling to the surface like froth. Five minutes ago, he would have disarmed at this comment, but not this time; running away is the mark of immaturity. He's an adult. He's gonna take a stand.

"I'll organize; I'll protest," he offers. "I'll use the pen over the sword. Maybe I'll run for office."

Pando scoffs brazenly. "Politics is broken," he sneers. "How can you trust any of those idiots to solve our problem? They've told us to go fuck ourselves. No one is gonna save us but us."

"Well maybe if people our age actually *voted*, our voice would have a little more traction," he challenges.

"Yeah right," says Pando. "We can't risk the fate of the world on the hope and prayer that our generation will suddenly wake up on our own. We need an alarm clock. And tomorrow, I'm gonna be that alarm clock. What about you?"

Cooper considers the gun in his hand. Raw power — the power to kill in an instant. No wonder humans are on top.

Pando, tracing Cooper's gaze, resumes his assault. "What's it gonna be, Coop? Are you gonna help me catalyze this revolution, or are you gonna kill it in its crib?"

Scared and mad, Cooper looks up at his friend, who also seems to be shaking. "No," he says firmly.

"No *what?*" Pando detonates.

Without words, Cooper hides away his gun and walks toward the exit.

"Cooper? Cooper!"

It's no use. The man has left.

Cooper's lively pace carries him quickly down the block. He can hear Pando shouting for him.

"Cooper!" his brother screams.

He breaks into a run, rounding a corner to elude his pursuer. The refugee sprints after him.

Panting heavily, Cooper looks for anywhere to hide. Pando, in superior form, is closing. Exhausted from the chase, Cooper's pace slows, his feet catching on the crumbling sidewalks. He ducks into a dim alley, hoping Pando may not notice. He finds him anyway.

"What are you doing?" demands Pando, cornering Cooper in the darkness.

"Not helping you," he states. "Don't you *see*, Pando? You can't go out like this."

"And what else am I gonna do?" he pleads.

"Your music, it's genius," Cooper offers. "It's more powerful than any bomb. You have a *gift*, man. Don't rob the world of that by killing yourself like this."

Pando pulls the gun out of his pocket. "Cooper, please..." he says angrily.

"I'm not doing it," Cooper resolves, ignoring the threat of the weapon. At last he has broken free from the shackles of fate, ready to carve out a new destiny.

Pando brings the gun up to his temple, the cold metal digging into his scalp.

"Then I may as well kill myself now," he quavers, "because without you I can't do this."

Fear grips Cooper's body, his amygdala lighting up like a meteor in the sky. But then he flexes his brainpower, choosing instead to think it through: would Pando really kill himself right now? No, he wouldn't, he reasons. He wouldn't throw away his life this uselessly. He still has a job to do. Cooper decides to call his bluff.

"Pando, put the gun down," he says with calm reassurance, his frontal lobe reigning in his riotous emotions. "You're being dramatic."

"Don't do this, Cooper," Pando begs, the gun still to his head. "Please. I need you."

The new man rises onto his toes and puts his face right in Pando's, their noses touching. The steel in his eyes is reflected in Cooper's. They flare like binary suns.

"I will oppose you," breathes Cooper. With that, he brushes Pando aside, then strides out of the alley and into the broad daylight beyond.

The refugee watches him leave. He didn't even look back.

"So be it," Pando whispers to himself. He draws the gun from his head and replaces it in his pocket. His nerves, though frayed, begin to collect themselves again.

It's late; he'll have to get back to BKP. He isn't looking forward to the impending conversation with Iri, where he'll have to break the frightful news: Cooper's gone rogue.

4C

The plush magenta cushion gives generously beneath Shree's knees. She kneels at the vacant pew, the empty church a vast, hollow chasm. She has come seeking salvation, for she knows nowhere else to turn. A Bible is flipped open beside her, the passage of Isaiah 24:4-6 marked in red highlighter. Her head is bowed in prayer.

"Dear Lord," she speaks quietly, "please show me that there's still a reason to have faith."

God seems to have dismissed her plea, the debilitating silence continuing unabated. It's as if He's turned his back on the world, or, more likely, the other way around.

She returns home later that warm Tuesday night, a fallen being in a fallen world. Upon entering the apartment, she is greeted by silence all around. Mel is at a bar tonight, drinking to excess as she rethinks her career plans. Cooper, having defected from BKP, is on the lam. And without Pando, she is completely, utterly alone.

She flops onto her bed and turns out the light. The darkness consumes the room, devouring it lip-smackingly. The blackness filters her thoughts.

Her class of 36 was too big to handle. Too many children, and not nearly enough time for all of them. Getting them to pass their exams was an impossible task. Was it her fault for letting them down? Or was it their parents' fault for bringing them into a world that doesn't care?

A flashback takes her to the gym, her students working endlessly as the seconds tick away. Now that she's gone, who's gonna teach them the science they need to know? Who's gonna teach them not to fuck up the world when climate change isn't even in the curriculum? They're like little parasites squirming in their seats, sweating crude, belching smoke, shitting bricks of coal, drooling gasoline. They're all polluters, all of them. So much carbon. So little time.

420 metric tons. That's what her life's been worth so far. Now multiply it by three, for an average lifespan. 1,260 metric tons. Altogether too high.

She opens up the carbon tracker webpage from the BKP presentation. There in red ink, stretched across the top of the screen, a counter tracks the tons of gas, its digits whirring upward in a muddying blur. Ten seconds later, another thousand tons have been released. The countdown to doomsday ticks, the seconds shearing off one by one, until the day, not far from now, when humanity's hourglass runs out.

She sits in her empty room, curled in a ball, staring at nothing but the light from the screen. Within minutes, she freefalls into a rich sleep.

Flashes of fear wring Cooper by the neck. He looks over his shoulder every few paces to make sure he isn't being followed. BKP could be anywhere. So could Cokeheads.

He's removed the battery from his phone, hoping it will prevent anyone from tracking him. It seems that on every street corner, he glimpses "Keep It in the Ground" fliers advertising for the protest, "#decarb" stickers, and the Brooklyn Project logo, placed by anonymous revolutionaries. He feels the influence of BKP everywhere, its spidery tendrils snaking through the city like vines.

He finds his way to Prospect Park under the cover of night, still unsure whether or not he's being followed. He heads for a low hassock of brush, taking a seat on the grassy carpet behind it. At least here he's out of sight, from cameras, Cokeheads, and BKP.

His body is quick, responsive, like it hasn't been all year. The hunger is gone, suppressed by the fight-or-flight response. The exhilaration of his escape has distracted him from the irrefutable truth, the driving force that has kept him on the move for the past few hours: he's going to die. But it's this very knowledge that empowers him to live. He's accepted his own death; now it's time to make something of the life he has left.

He can't save the glaciers. He can't save the oceans. He can't save the rainforests. But maybe, just maybe, he can save people. And maybe that's good enough.

It's certainly good enough for Pando, who feels that he's about to do the most good for the most people. In the hours since Cooper abandoned him, he has regrouped with Iri and updated their plan of action. She will replace Cooper for the mission. Since the rogue can bring all of them down if he says anything to anyone, they've dispatched over a dozen agents to comb the city and capture him. Pando doesn't know who any of them are.

He checks the time on his phone; it's 11:50pm. In nine short hours, he'll make his way to the parade ground as one of the first protesters, trigger in hand. In twelve, he'll be dead.

He sits alone at BKP, in the same presentation room where his life changed on that momentous night five months ago. He's bunking here tonight, since tomorrow is the big op. Not that he would have anywhere else to go — he's been homeless since Shree broke up with him.

Iri is working with another agent to scope out the Columbia campus, assessing where they'll put the extra security now that Cooper's a random element. They'll have to surround the area in a virtual blockade if they're gonna get enough eyes on the ground to keep him out. That is, if he even decides to come at all.

I will oppose you, he had said. He'll be there. He promised.

Pando looks up at 22nd century Hell. It taunts him from the wall, daring him to make a move. Not wishing to spend his last night on Earth in the back room of a bar, he takes up that dare. He heads outside for a walk, to appreciate for one last time the smooth air of night, black like oil.

He looks around at the other bars along the street, buzzing with people gathered for drinks, cackling like hyenas. They all look like they're having so much fun, enjoying life in a way Pando hasn't since Typhoon Powell crashed into his life. In fact, he may have forgotten how to laugh completely at this point. Is this why he's so ready to kill himself? Has he in fact been dead this whole time? Did he die in Kabubwason with his brother? Has it all been a nightmare unreality?

The humans are out in force tonight. They're all so carefree, so breezy, watery smiles on every face. From the looks of them, you would never know this is a civilization in collapse. In fact, on this warm night of pre-summer, things look rather idyllic. Even Pando almost buys into it. As he gazes at the Manhattan skyline across the river, he feels a blush of hope: maybe people aren't all that bad after all. If we can live lives this rich, this full, this transcendent, then we just might make it. Yes, we might.

He walks down to a stretch of river and leans himself against the rusted iron railing. The skyline glitters like quartz before him, a sprinkle of stars peeking through the muted, smoldering sky. The imperial reach of mankind stretches far above, drowning the atmosphere in a blanket of methane and soot.

His hand reaches into the pocket of his shorts, pulling out a crinkled photograph. It's peeling at the edges, and appears to have been bloated with seawater at least once in its life. It's a picture of his brother.

"This is for you, Lito," he says softly, looking into his brother's eyes one last time.

He flips the photo over, where, sketched on its jaundiced backside, there is a passage from the Book of Revelation. He wrote it there the day he accepted his mission, as a reminder that sometimes religion gets things right:

> The nations were angry; and Your wrath has come. The time has come for judging the dead, and for rewarding Your servants, the

> prophets and Your saints, and those who reverence Your name, both small and great — and for destroying those who destroy the Earth.

He clasps his hands and bows his head, taking advantage of his last opportunity to give God a try. With the cloudy water rushing by beneath him, he whispers a prayer into the midnight air.

"Dear Lord," he says, offering himself sincerely to the divine. "I hope you'll forgive me for my sins, and for the sin I'm about to commit. I think you'll understand why." He sniffles loudly, searching for what else to say.

"Please keep Shree safe," he adds. "Keep them all safe. And if Lito is up there, keep him safe." He has no misconceptions of which place he'll wind up in the afterlife, if there is one. But hey, maybe Hell won't be so bad.

"And finally, thank you for the life you gave me, and the world you gave us. I'm sorry we ruined it." He shuffles his feet into the gravel. "Amen."

He opens his eyes, and releases the photograph from his fingers. It zigzags into the water like a leaf in autumn.

Just for good measure, he throws his head up and picks a star to wish on, in case the prayer doesn't make it all the way to God. He might as well try all methods of superstition while he's at it.

"I wish I didn't have to do this," he says to the tiny light in the sky.

In Prospect Park, Cooper is staring at the same star. How long has that star been there, he wonders? How many civilizations has it seen come and go? How many species?

Cooper thinks about those species, the ones that are already gone. His mind takes him to a heaven for the dead forms of life, a fictional world shrouded in linty clouds, where all the species that no longer walk the Earth get to live in perpetual harmony. There are so many, in so many varied and splendid forms, in shades of marigold, fuchsia, and olive drab.

There's the American cheetah, and the American lion along with it. The wooly mammoth and the moa. The sloth lemur and the quagga, Stellar's sea cow and the California bear. The dodo and the great auk, the golden toad and the bulldog rat. The sea mink, the passenger pigeon, the tarpan, and the Texas wolf. And those are just the ones with names.

Far in the distance, standing on a knoll, he can make out the shape of one more: a biped with opposable thumbs.

. . .

Dawn has arrived, the morning chorus chirping outside. Not from songbirds, but sirens, children, and honking cars. The city that never sleeps is waking up, and Shree along with it. The only light in her room is her monitor, still loaded to the carbon pollution webpage. It kept count all through the night; it's noticeably higher than when she last saw it. Her stomach wriggles nauseatingly.

She switches to a news webpage, deeming it better than watching the world suffocate. The first thing she sees: "Antarctica Doomed, Melting 'Appears Unstoppable'." The last leg of her mental stability finally gives way, folding like an accordion. Game over.

No more Antarctica, she thinks. A world without ice for her students. A world without ice for her. A world without ice is no world at all.

She rises from her bed and grabs a pen and sheet of paper. She works ink onto the page, trying to form letters with a shuddering hand. Once finished, she leaves the note on her bed.

With heavy feet, she makes her way to Mel's door, knocking on it thrice. "Mel?" she calls. "Are you in there?"

On the other side of the partition, a passed-out Mel doesn't stir from her bed. Convinced she never made it home last night, Shree gives up knocking. Her heart sinks; she would have liked to say goodbye. Instead, she grabs her phone wallet and exits the apartment, quiet as a ghost. Her keys remain on the hook.

In Prospect Park, the morning noise stirs Cooper as well — but the revelers are songbirds, and he hasn't slept. He's been keeping a scrupulous watch for antagonists. But the sun has risen appreciably into the smoggy air, spilling a milky pool of light into the city through a yellow sky. Even without checking his phone, he can tell it's already mid-morning. He'll have to get a move on.

Okay, okay, he assures himself as he hides beneath the cover of a low maple. *Just focus, okay? Relax. You're an adult. You got this.*

He knows he can't stop Pando alone; he has to go back to Bushwick to get Shree, for she's the only one who can convince the bomber not to go through with his plan. Fully aware that there are probably agents of one faction or another waiting to apprehend him at his stoop, he'll have to get creative if he's gonna get inside.

He gets off the train warily, his eyes wide like golf balls in their sockets. He walks prudently down the street, pupils darting everywhere, scrutinizing every face he passes. To an outside observer, he looks a paranoid schizophrenic. His hand is wrapped carefully around the loaded gun in his pocket.

A block from his apartment, he peeks down the street from around a corner of bricks to see if anyone is loitering nearby. Sure enough, he spies a man leaning against the side of a building not far from his stoop, standing with no obvious purpose. Not smoking. Not listening to music. Just chilling. The man is young, 20-something, dressed like a hipster. Cooper doesn't recognize him, but knows he's probably with BKP.

What to do, what to do, he thinks. The sun is even higher now; he has to get inside. He ducks behind the wall and looks around for witnesses. A couple is walking away from him down the street, but they hang a left, leaving the block deserted. Cooper draws his gun.

Plugging his left ear with his forefinger, he fires a shot straight up, its thunderclap reverberating off the buildings and down the avenue. Peeking around the corner, he sees that the unidentified man has turned his attention toward Cooper's intersection.

He fires another shot into the air, then a third right afterward. The man reacts now, and begins walking down the road in his direction.

He puts the gun in his pocket and sprints to the next street. He emerges on the opposite side of his block, looking back at the stoop from the other end. The man has gone, drawn away by the gunfire. Cooper dashes into the building and scampers up the stairs to apartment 4G.

"Shree!" he screams in desperation. "Shree!"

He scrambles up the rope ladder and starts pounding on her door. "Hello?" he shouts. "Anyone home?"

Why is he respecting her privacy at a time like this anyway? Forget knocking; he barges in. He flicks on the light, seeing nothing but an empty room.

Panicked, he gives Mel's room a try, pummeling the door with his fist before muscling it open.

"Hey!" the sleepy girl groans from bed. "What gives?"

"I need you right now," he says. "Get dressed. We gotta go."

She stares at him from her bed. The hysteria in his voice is evident; if she wasn't so hungover, she'd probably get up immediately. But in her current state, the warm embrace of her mattress is too sweet to give up.

"Why?" she cries.

Cooper considers the right words. He doesn't want to reveal anything incriminating; if he says too much, BKP and everything it stands for will be for naught.

"I think Pando's about to do something terrible," he says.

Mel frowns. "Who gives a shit about him," she mutters, turning onto her side.

"This isn't about him. Get up."

She looks up from her pillow. "Why, so we can 'save the world' or whatever?"

The blood scalds Cooper's face. There's no time for this bullshit.

"Mel!" he roars, reaching a volume that shakes his roommate upright; the man she helped unlock within him has her full attention. "This isn't some fucking joke! I need you! I need you right now! So shut the fuck up and get dressed!"

The girl blinks in bed. After a moment of silence, she grimaces skeptically, then turns to roll onto her feet. "This better be good."

In a frenzy, Cooper rushes to his room and looks around for anything that might be useful. Anything but the gun in his pocket.

He digs through his duffel, slinging apparel through the air like a monkey tossing its shit. He finds a blank white shirt suitable for his purposes. He locates a pen on his table and stabs a hole through the cotton breast, which he tears open to enlarge into an eyehole. He rips another a few inches to the left, then draws a large black nose beneath their midpoint. From under it he shapes a clumsy muzzle, giving the vague impression of a wild animal, drawn with the artistic prowess of a four-year-old.

"Good enough," he pants, springing to his feet. He lurches for the door, but something on the table catches his eye. It's a rubber band.

Cooper's mind crafts a bold but necessary plan. He snatches the little ring and runs over to Mel's door.

"You ready?" he asks, watching her pull on a shirt. He grabs her by the wrist and yanks her onto the rickety walkway.

"Calm down!" she flares.

"Then hurry the fuck up!"

He jumps the last few rungs of the ladder, abbreviating his departure by less than a second. But that fraction of a second could be the difference between life and death.

"Listen," he says before they leave the safety of closed doors. "There might be a guy out by our stoop. I need you to distract him so I can leave without him seeing me."

He doesn't explain why, nor does Mel ask. They race down the stairs until they reach the front. Through the glass in the door, they see that the hipster man has returned. He is stationed directly on their stoop, sitting on the steps and overlooking the street. He's waiting.

"That's the one," Cooper whispers from behind cover.

Mel nods curtly. "I know what to do. Just wait for the right moment."

"Alright. Meet at the L train. If you're not there in five minutes, I'm going without you."

Her eyes glow with a certain human understanding. "I'll be there," she promises. He believes her.

The door creaks open; the agent looks aft. Mel walks with slovenly grace as she passes in front of him and to the foot of the stoop.

"Hey," she says with a voice sweeter than chocolate to whet his appetite. "Do you want to get in?"

The young man's face blushes. "Oh, uh, well I'm actually waiting for someone," he says, his eyes scanning her over.

She considers him conspicuously, making it clear she's checking him out. A step forward signals her interest, her eyes smiling with intent. "Are you bored?" she asks.

He looks around bemusedly. "Kinda, I guess. Why? What's your name?"

She comes in close to him, then lifts up her shirt, offering her bosom for his slender hands. "What about now?" she whispers.

Cooper, watching the scene from inside, recognizes an opportunity. He wraps his fingers firmly around the cylinder of metal in his pants.

Before the horny man can lay a finger on Mel's sultry white breasts, she lets down her shirt. "Sorry, not this time," she says. With the power of an iron club, she thrusts a dire blow into his unsuspecting nose, crushing it like a walnut.

Cooper throws open the door, brandishing his deadly weapon. The bloodied man jumps to his feet, wheeling round to defend himself, but not before Cooper lands a punishing strike on his cheekbone, smacking him with the handle of his gun. The man is sent to the ground.

With feverish diligence, Mel scans the road for any witnesses. Cooper leans low to assess the condition of their victim.

"Is he gonna be okay?" she chuffs.

Cooper listens for a pulse, gauging the disabled man's soft breathing. "I think so," he gasps. His hand infiltrates the man's pockets, raiding them for his phone. He opens the casing and yanks out the battery, replacing the dead hull back in the appropriate pocket.

"Run!" he commands, battery in hand. The two of them gallop away down the street.

The subway ride is sinisterly long, punctuated by extended periods of idling due to train delays. Cooper's stomach quivers like an egg white. Mel, fully alert, looks ready for armed battle.

"What's that in your hands, anyway?" she asks him, pointing to his white shirt.

"Oh. Polar bear mask." He pulls it over his head, discovering that he made the peepholes too close together, giving him a cross-eyed look.

She nods stiffly. "Nice."

They depart at 116 Street and hike up the stairs into the sun. The noise of a large gathering is readily audible in the air, chants of rebellion echoing through the atmosphere.

He can feel it growing as they approach. It throbs through his soul, and everyone else's. Change is percolating, about to burst forth onto the streets of this nation — a deluge to thrash against the crumbling levees until it floods the plain. A revolutionary tension in the air, taut as a noose. The energy is ready to be flared from the system like methane, a powder keg awaiting its spark. From all angles, the mood is positively electric, like the epicenter of a movement — or the crater of a storm.

From this moment, at this rally, the crusade will either live or die. And humanity along with it.

He isn't the only protester sporting an animal suit today; he hardly looks out of place. Behind the security of his mask, Cooper passes effortlessly into the field of people, undetected by the BKP agents spaced regularly around the ever-enlarging perimeter of the crowd. There are clearly far more people in attendance than anyone had anticipated.

"Keep it in the ground!" chants a pocket of protesters, hoisting their picket signs high into the air in a rhythmic mantra. "Keep it in the ground!"

Cooper observes the mélange of people, a throng of students and adults alike, the full spectrum represented across their infinite styles of hair, clothing, jewelry, and costumes. The creativity is overwhelming; how fiercely individual they all look, yet completely indivisible in spirit. Tropical house music plays over the speakers on the stage, the scene more akin to a rave than a protest.

An active rectangle of people fills the walkway down the South Field, corralled by the green hedges to either side of the path. Their numbers spill back past the sundial and onto the perpendicular College Walk, with more people amassing at the fringes. The stage has been erected down by the stairs of the Butler Library, just as expected. The only problem is that five thousand people stand between here and there.

"I've got to get to the stage," he shouts to Mel. "You find Pando!"

The pair split, Cooper making a B-line for the side entrance to the library. As he approaches, he notes that the stage has been offset from the mob by a thin line of pink tape, creating a ten-foot-wide path between the wall of protesters and the front edge of the platform. Tied to its floor is a hanging banner that covers the gap between the metal scaffolding and the shadowed ground below it. Somewhere in that space, Cooper knows, is an IED.

Mel shoves her way through the crowd, her eyes sweeping back and forth thoroughly. Pando is here, most likely at the front, and she will find him, though she still doesn't know why. Pushing into the throng of screaming bodies aggravates her hangover, but she ignores it gallantly. She has a job to do.

Cooper makes it to the west wall of the building, looking around the corner to weigh his approach. He can almost see underneath the stage through the side stairs, but it's too dark to make out anything for sure. With a direct approach out of the question, he slips into the library, quiet as a mouse.

The jostling of the crowd will leave Mel with more than a few bruises by the time this is all over, but she continues her slow advance to the front. Her height provides an important advantage, climbing onto her tippy toes to stop and scan every now and then. Now twenty feet from the front, she notices a familiar knot in black hair. It's Pando.

He stands directly before center stage, his body even with the podium as the pink tape presses into his chest. He is as close as he can be to the weapon under his control, the thin blue banner the only obstacle in his line of sight to the bomb. There's no way he's out of range from here, and no way he'll survive.

With green warpaint slathered on her face, Iri stands further to the left of the crowd, ready to step in if Pando is out of position when the moment comes. Her emergency trigger sits wrapped in her pocket. The smartwatch on her wrist sends data to BKP, where the Sisters are monitoring the op. She searches the crowd diligently for any sign of Cooper.

On each side of the stage are two of her agents, hand-selected for this mission. With official Columbia lanyards around their necks, they look like student organizers, and indeed they are. What sets them apart is that they're also with BKP, and are using their in to keep an eye on the crowd for any rogue agents.

"Nothing yet," Iri says into her headset. Pando barely hears it through the tiny speaker in his ear, not quite powerful enough to overcome the chanting of "Decarb or die!" sounding about him.

"Pando!" cuts a voice from behind; he turns around to see. Through the heads and signs of the crowd, he glimpses Mel, their eyes making contact. She looks like a predator, the hunt in her stare. He immediately retreats into the crowd.

"Pando!" she shouts again, losing him in the mass of bodies. Furious, she burrows deeper, determined to subdue him.

He muscles his way toward the middle of the gathering, hoping to circle back to the front by the time the speaker takes the stage. Until then, he'll have to evade.

"I'm being tracked," he says into the mic.

Iri leans her head over the tape, noting that Pando has left his station.

"Where are you?" she demands.

"I dipped into the crowd. I'm gonna make my way back in a few."

Cooper emerges through the library front entrance, the stage now directly before him. Its rear wall, covered with signs that read "#decarb" and other mottos, provides an effective cover from all but the extreme edges of the crowd. To his right, he sees a white-roofed tent with a number of what look like organizers. One of them is Natasha.

With a dash, he flings himself under the stage, praying no one has noticed his approach. He rolls over the scraping cement, stopping on his back. Directly above him, adhered to the bottom of the platform, is a suitcase-sized device, its red eye winking at him ominously.

"Oh shit," he says, face-to-face with an armed bomb for the first time in his life. He pulls off the polar bear mask to get a better look. It's bigger than he imagined; this might not work after all. Maybe he can just move it somewhere safer, away from the stage. He tries to pry it off the underside of the platform, but it's no use; it appears stuck. He tries again with all his strength, until he realizes that it's physically affixed to the floor by two pairs of screws. So much for that.

The volume of the crowd drowns out all coherence of thought. Dimly he studies the box, running his fingertips over a latch on its side. He hinges open the casing, revealing the deadly mechanism within.

It looks nothing like the schematic he remembers from Iri's briefing, if his foggy, sleep-deprived brain can accurately recall any details at all. All he knows is there's some sort of switch that, when released, will lead to detonation. Now to find the switch.

Behind the banner, the roar of the crowd swells into applause. As he fiddles with the device, Cooper sees the wooden floor shifting under the weight of shuffling feet. Someone has taken the stage.

"Good morning, ladies and gentlemen!" says a creamy voice. Cooper's heart pounds faster; Natasha is right above him. Time's up.

Pando, still hiding in the crowd, turns back for the front again. Mel, meanwhile, continues to search fruitlessly. She's lost him.

"Fuck," she mutters. The crowd is too dense; he slipped away like an apparition.

"We're here today," says the voice from the speakers, "to demand our future back from the carbon polluters." The crowd goes wild, banners waving madly in the air, the ground rattling seismically from the stamping of their feet.

Cooper stares at the bomb, searching for anything that looks like the switch in question. He notes a small metal rod in the middle of the device, clearly angled away from a rather important-looking tangle of wires, but he's hardly certain this is conclusively what he's looking for.

He reaches into his pocket for the rubber band, his plan B. With no other leads to go on, he carefully wraps the ring around the metallic switch, praying to God that it might disable this weapon. He ties it satisfactorily, careful not to touch any other part of this crudely-constructed explosive. He removes his hand slowly; the band stays. He takes a deep breath.

Now to get out of here, he thinks.

"We're going to send a message to the leaders of Columbia University that investing in fossil fuels is investing in evil," Natasha bellows into the mic, sending the crowd howling into a frenzy. "As long as the carbon polluters have our money, they own our future. They're like bad parents, throwing their babies into the oven, because being responsible adults is just too damn hard for them."

Cooper wriggles his way under the back lip of the stage, but the turbulence of the ground catches him by surprise. The crowd is eating up Natasha's words. From out of the corner of his eye, he notices something drop from the blinking bomb and onto the ground. It's the rubber band.

Pando has returned to the front of the crowd again, working his way toward the center. He hears Iri over the radio: "Wait till she's finished speaking." He has no time to lose.

Foiled by the rocking passion of the mob, Cooper picks up the fallen rubber band and tries affixing it again. He wraps it around twice this time, hoping the extra tension will keep it in place. The energy of the crowd, meanwhile, is reaching a boil.

"Why are we letting them run our lives?" Natasha asks, riling her audience. Every angry student is consummately amped, the heat and energy pouring off enough to melt glaciers. "When is our generation gonna wake up? Their dream isn't our dream. When will we finally stop chasing it?"

Pando is back in position now, frontmost-centermost as planned.

"Everyone, pull out your phones right now," demands the leader on the stage. "Tweet '#decarb' for all the world to hear!"

Mel, lost in a sea of riotous protestors, can do nothing more. She's lost Pando, and too far from the stage to see him.

"Hashtag decarb!" scream the youth, phones in the air. "Hashtag decarb!"

The rubber band flings loose again, shooting into Cooper's face. It isn't going to hold, and he's out of time. The bomb could go off any second.

"We'll show the establishment that its ideas are old, that if they won't do the right thing then they should stand aside and let us do it," Natasha blisters. "There's no reason we should suffer for their mistakes."

What to do what to do what to do, darts Cooper's mind. He can't disable the bomb. What would a man do in this situation, the woman he cares about in deadly danger? He'd take the bullet.

Once more he wraps the rubber band around the lever, hoping it'll at least buy him time. Maybe, if he can calm the crowd, it will.

"They're too old to adapt to the cunning pace of change," the speakers blare from the sides of the stage. "But we're young, fast, and clever. We'll race ahead while the wrong side of history devours them like a hungry polar bear."

Cooper scampers out from under the stage.

"Raise your hand if you're an alarm clock," Natasha says to the crowd, "an agent of change looking to wake up the world." Hundreds of hands immediately shoot up before her, those not realizing the deeper meaning following suit after a moment of hesitation.

"We're gonna take back our future whether the geriatrics in power want us to or not," she cries. "We're gonna scream at the top of our lungs that burning fossil fuels is a crime against humanity. Who's with me?"

Before the audience has a chance to react, Cooper, with foolhardy courage, enters stage left. He strides up to Natasha with calm docility, slipping the microphone right out of her hand. Dumbstruck, she takes no action in protest; in actuality, she's quite happy to see him, and thoroughly alarmed.

"She's right," he hollers to the crowd, quelling his anxieties for the first time in fifteen years. All heads strain to see this new voice. Pando, directly in front of him, is frozen in utter shock. This absolutely is not the plan.

"Dammit Cooper, you fucking idiot," Pando growls to himself. "Don't make me do this."

The man stands proudly on the stage, his stance tall and his chest wide. "If we're gonna take our future back, we need to take this fight to every corner of the globe," he shouts. "And if the people in charge of the United States of America refuse to do the right thing, then we will remove those people from the seats of power and start anew."

The rowdy morass seethes with conviction, guzzling his words like liquid courage. The rage is intoxicating, the singularity of purpose divine. They're a herd of aurochs, ready to charge. The rubber band underneath Cooper's feet has undoubtedly given way for good; an armed bomb awaits further instructions.

Natasha takes a step back, fully yielding the mic to the young prophet beside her. He motions for her to leave the stage, but she folds her arms and shakes her head no; she's waiting to hear what else he has to say.

His words upon taking the mic were just verbal vomit, clickbait for the headlines. He knew he had to tell them what they want to hear, but there's no courage in that. If he's gonna save Natasha's life, he'll have to disarm them all. Even Pando.

"Wait," he says, "wait hold on, stop. Everybody stop screaming for a moment. I want your attention."

He crouches low to the stage, leaning over the edge to get as close as he can to his captives. Their voices have mellowed, the earthquake dissipating. "I take that all back. We're not going to tear down this country."

The people, mostly silent now, listen to this with confusion. Pando squints skeptically, his heart thrashing in his chest. A quiver assaults his body, the no-win scenario revealing itself again.

"We aren't here to bring about the end of America," Cooper avows. "We're here to save her. We're here to restore justice to her land and strength to her soul. This is not an armed insurrection. This is a peaceful revolution."

He hears a whoop from the audience, then a smattering of applause.

"We will fulfill our duties as good citizens and make this country great again." He looks out to the crowd, noticing Pando's glinting chrome eyes. Cooper smiles confidently. That's right. He's got this.

"So put down your guns, and your hatred, and your fear," he calls. "Don't succumb to the habits of deniers. Let's put our heads to the problem. We don't need to change the world. We just need to change who we are. If we do that, the rest will follow."

He pauses for dramatic effect, scanning the audience like an eagle. "You want to solve climate change? Then get off your asses and do it. Don't just tweet some stupid hashtag and call it a day. Put your screens away and get a fresh breath of reality. The future is calling us, and we need to be paying attention."

Cooper notices the phones in the crowd dip slightly at this charge. Many of them now feel awkward holding up their electronic slates, witnessing this moment in history through the pixelated camera of a phone, rather than with their flesh and blood eyes. Still, this is a sermon that simply can't pass by without being documented by any fewer than five hundred vantage points. They owe it to history to keep recording.

Iri watches in the corner of the crowd as Cooper steers the ship from the stage; her eyes are trained on Pando. "The moment he's done, you fire," she commands.

Sweat pours from Pando's face, the internal struggle duking it out Thunderdome-style. He sees his brother on stage, so close. The waters surge all around, the biblical rain drowning the land. But look at the children. He can still save them.

"Pando!" shouts his commander. "Do you copy?"

The refugee snaps to, swallowing hard into the mic. "Copy."

Mel, in another part of the crowd, observes Cooper's hijacking from afar, mouth agape.

Cooper doesn't see three thousand people before him; he sees one creature, a singular pillar of humanity. He speaks to the whole, as part of it.

"We will stand up as one, and we will prove to the universe that we're all grown up." He looks up to the carbon-infested sky. "We will fight the deniers who seek to enslave us, those too old in their thinking to see the clear signs of change coming their way. We are that change. And we don't have to listen to them. We will ignore their voices, and embrace the voice of reason: the voice that God gave us. In billions of years across millions of species, we are the first and only ones to wield the power of intelligence. And look at what we do with it. We fly airplanes. We cure diseases. We travel to other worlds. We are gods. And we're smart enough to get through this.

"We all just need to focus. Focus and relax. Take a deep breath with me."

He inhales deeply through the mic, humanity breathing with him. The oxygen feeds their brains, and brings a moment of clarity to everyone's heads.

"See?" he calms. "Just relax. We're human. We got this."

He looks at Natasha, a smile on her face. Then he looks out to the enraptured congregation before him. Even Pando seems enthralled.

"Let's get to work," he says finally. And with that, he throws his arms in the air, the crowd convulsing into booming applause.

Pando weighs the words of Cooper's speech. *Just words,* he thinks. *But please. Get off the stage.* He glances over at Iri, who's leering at him, waiting. It's now or never.

He takes a deep breath. If Cooper wants to get in the way, then so be it. He's sacrificed his brother once before; it was the right choice then and it's the right one now. Feeling the pull of destiny, he slowly reaches into his pocket, fishes out the slender rod, removes the protective lid, and brings his thumb to the button. He pauses, taking a long look at Cooper, his brother in arms.

He was so young. We all were.

He closes his eyes tightly, breathing in sharply through his nose. With clenched jaw, he mashes his thumb onto the trigger with the force of an atom bomb.

Save Me Now

Enough is enough has long since passed. In the shaking silver tube, Shree's mind has disengaged entirely from her surroundings. Her unconscious is in command, her awareness enmeshed in a web of painful memories. She can't shake loose the terror of everything she's experienced during her brief foray into the real adult world.

The skies turn a sour black, Apocalypse unleashed upon the Earth.

What was it that did her in for good? Was it teaching in the sweaty gym, brownouts cutting power to the fans every two weeks?

A deluge overwhelms the seawalls, the city streets licked clean by tongues of ocean.

Was it watching her students fail, no matter how hard she tried to make them understand?

Boats wash ashore like child's toys in a bathtub, hurtling into buildings knee-deep in standing water.

Was it fudging the test results, the final indignity to the institution of American education?

The city is subsumed by the sea, her body along with it, the inevitable finally coming to pass. Judgment Day.

She emerges at 168 Street; the legs that carry her along are not under her command. She walks the city blocks to the bridge: the tried-and-true final solution, the surest way to end the pain. She shuffles along absently, knowing she has already made her final choice, the choice she was always going to make, the choice she somehow knew she would someday make. She has embraced her fate, making her escape from the prison of this tyrannical existence the only way she knows how.

As she climbs the rail to overlook the Hudson below, she takes a final drink of the vista in the distance, the glass pillars stabbing into the peculiarly fire-red sky. She ignores the uproar of the cars, the smell of asphalt and exhaust, and focuses on the cool breeze of God's breath flowing over her along the river beneath.

In a moment of clarity, she slips her hand into her pocket. She pulls out her phone and dials a familiar number.

It rings once, twice, three times. After five tones, it goes to a recording: "Hey it's Pando, text me instead. You know I don't check voicemail."

Her hands begin to convulse violently, fear seizing her muscles as she fights the urge to die. On the post beside her is a sign with reassuring words: "You're Not Alone. Call Us. Suicide Hotline."

She begins to type the numbers onto her screen. Suddenly an unexpectedly strong gust of air catches her body, blowing her against the railing. Her muscles tense as she wraps her arm around the cold metal pole, her shaky hand losing its grip. The phone tumbles out of her palm, somersaulting into the tributary with a white splash.

Nervous laughter overcomes her, relieved to finally be out of options. The universe is telling her that now is the time.

She stares into the green-gray waters rushing beneath, soured by the filth of humankind. Will it be enough to wash away her sins? Only one way to find out.

She feels a release from deep within her soul, the knot finally coming untied. Her fear slumps away, replaced by surging elation and confidence of purpose. In the distance she can make out the face of her Creator, smiling down on her for one last time.

A deep breath later, she jumps into the air with conviction, the river below lurching toward her. With a splash muted by the drone of midday traffic, the 23-year-old returns her body to the Earth, her spirit rippling through the air like a comet, bound to meet her Maker.

Five kilometers to the south, Cooper finds himself standing before an incanting crowd, the chorus of screams loud as a nuclear explosion in his ears.

Natasha claps wildly off to his right. She steps to his side and places a gentle hand on his shoulder. "Good job," she whispers.

She wrests the mic from his grip as he stands frozen in a perplexing state of elation and terror, the power of his oratory a dangerous and frightening new gift, strong enough perhaps to disarm the bomb blinking under his feet. For still he stands, still he breathes.

Natasha's voice comes through the speakers. "You heard him everyone! This is our generation's fight, and all of you are part of it! Thank you for coming out! We'll stay all day! We'll take this to the top! And we'll win!" She grabs Cooper's arm and they begin walking off the stage. Cooper awaits the explosion that's about to change the course of history.

But it doesn't come.

The two of them dismount the platform, waving to the crowd as they make their way to the booth off to the side of the setup.

Pando looks at Iri with an expression of complete shock on his face. He pushed the trigger. Nothing happened.

Iri's searing eyes say it all: he failed. She scowls ferociously and begins moving through the crowd to meet him. He follows suit, pushing people aside as they chant, "We are the change! We are the change!"

She pushes her emergency trigger furiously, the bomb obstinately refusing to detonate. They meet close to the edge of the mob.

"I don't know—"

"Just forget it!" she shouts. "We gotta abort."

"But what about—"

"Go!" she commands.

Pando blinks once, struck dumb by this catastrophic malfunction — just like that, a revolution is snuffed out like a candle in the breeze. He turns numbly, sloshing through the bodies and on toward the columns of the Memorial Library up the hill.

Iri watches him disappear into a sea of people. She starts for the front again.

Cooper watches the crowd with throbbing exhilaration. Before him stands a monolith of humanity, singularly captivated by the words expelled from his mouth, unloosed before an active audience for the very first time. Posters and banners rah deliriously above them, smartphones poking from the crowd like ostrich heads.

The Clean Front leader angrily wriggles into the nonexistent gaps of the shoulder-to-shoulder crowd. She begins chanting along with the rest of them, so as not to draw suspicion as she forces herself forward. After great effort, she thrusts herself into the front row, nothing but pink tape between her and the stage.

With broad gesticulations, she grabs the attention of the two guards planted beside the banner hiding the bomb, the agents whom she had instructed not to interfere with Cooper's unscheduled speech. She brings her left hand to her nose, giving it but the slightest scratch, and returns it to her side. She looks to her two cronies, sees them close their eyes for a moment, and watches as they turn their backs to the crowd. Her message has been acknowledged.

The two women bend down and reach under their respective set of steps at the sides of the stage. They each grab a wooden dowel featuring a round hole at one end. They move toward the corners of the banner, unknotting the twine that affixes it to the structural skeleton of the platform. The women tie the banner to their dowels, raising it high above their heads as they too join the crowd in its cheers.

The exposed belly of the platform is now visible to all, and as the women begin moving off to the right of the protesters, Iri is able to see the faint red blinking of the device which she had so carefully installed beneath the podium only a few hours earlier.

The people around her are so caught up in the energy of the scene that no one else has discovered the mysterious object clinging to the underside of the gray metal stage. Iri looks to her right, examining the man beside her. He is tall and muscular, with a red bandana tied over a shaved head. The tattoo on his arm reads "Freedom" in red and blue stripes. She considers him carefully, before turning to her left instead.

On the other side stands a tall, thin black girl, by the looks of her a freshman or sophomore, with a cardboard sign in her hand that reads "Love Without Cause" waving up and down. Iri nudges her ribs.

"Hey," Iri speaks, "do you see that thing under there?" She gestures directly to the bomb with a sweep of her index finger. "That little blinking light? What is that?"

The girl lowers her gaze to see what she's pointing at. "Uhh..." she drawls, squinting her eyes.

"Do you see it?" Iri asks again. The stage isn't more than ten feet in front of them, and the girl can clearly make out the pulsing red light.

"Yeah, I see it. Huh. I dunno what it is."

Cooper and Natasha join the other organizers in the campaign booth. He feels a vibrancy within him that he's never known before. He has tasted power for the first time, and it stirs his every tissue. Not only did he steal the show, but he saved the day, and Natasha's life.

Suddenly he feels a tight hand clutch his bicep, yanking him to his right with a force that threatens to dislocate his arm.

"What was that?" Natasha asks, pulling him behind a broad brown trunk and shaking him from both of his arms. Her cutting tone, along with the throbbing pain in his shoulder, sweeps away any feeling of invincibility that he may have momentarily experienced, a sharp reminder of his true mortality.

"I'm sorry," he beseeches. "I had to do it. I'll explain it all, I promise."

She frowns with damning eyes. "Okay, so explain it," she rages. "Who do you think you are?" Her eyes seethe an anger Cooper has never seen before, and he withers at the thought of her hating him for saving her life.

"You don't understand," he begins, fixing his shining eyes on hers. She breaks her enraged stare to peek around the side of the tree, glancing toward the crowd. She makes note of the hoisted banner now encircling the students, and returns her attention to the frightened man in her grip. Relief overcomes.

"Well it worked," she mutters. "They fucking loved it."

Cooper exhales loudly, a weak smile creeping onto his face. She smiles back at him, closing her eyes as she brings her lips to his cheek for a peck. "Thank you," she whispers. His face flushes with warmth, his heart melting into a putty.

Iri ducks under the pink tape that holds back the crowd, approaching the stage with feigned curiosity. She pokes her head underneath the platform and looks straight at the deadly weapon, the blinking red indicator announcing its presence before the crowd. She considers with precision how long she shall dwell on the object. *One Mississippi...two Mississippi...long enough.*

"It looks like a..." she says, turning back toward the young woman in the crowd. "It looks like a bomb!" She runs forward, screaming, "It's a bomb! Bomb!" She waves her arms hysterically, drawing as much attention as possible toward the IED.

"Bomb! Bomb!" she shouts again, quieting the chants at the front of the crowd, now heeding the implication of her cry.

Cooper of course hears this, and his eyes shoot wide open upon remembering that lives are still in danger. Natasha cocks her head back toward the crowd. Releasing him from against the tree, they whirl around to see the commotion.

A scramble has broken out among the protesters, people moving in every direction. Cries of "Bomb!" spread through the crowd like fire through kindling. The chanting ceases, devolving into chaos. Terrified students drop their banners and noisemakers, pushing and shoving their way to safety.

Cooper gasps. He turns to face Natasha, only to find she's already sprinting away. A stampede of bodies threatens to crush his meager frame as they peel past him from the stage. He bolts from the scene, hoping to catch up to her, but it's too late; she's gone.

"Shit," he rasps, following the hordes to the nearest street and away from danger. *Where's Mel?* he wonders to himself. He puts the thought aside, knowing it only in his power now to save himself.

He makes it to Amsterdam Ave and across the street, where the wave of charging pedestrians has blocked traffic. People slow to a jog as they catch their breath and collect their thoughts.

Cooper pants in place as he turns back toward the scene he just fled. He pulls out his phone to replace the battery, then texts Natasha, Mel, and Pando with twitchy fingers:

Cooper: Where are you??? 11:58am

People continue to flush from campus, police officers rushing in against the current. Within minutes, a helicopter is hovering overhead, the beating of its wings eclipsed in tempo only by Cooper's palpitating heart.

Mel replies to his text, a brief report of her condition and whereabouts. They meet up minutes later at Straunt, sharing a warm embrace upon sight.

"Holy shit, that was fucking crazy," Mel says.

"Did you find Pando?"

"Yeah, I was trying to follow him when everyone started running. They said there was a bomb!"

Cooper lets out a long breath, happy to know Pando is still safe. "Yeah, it was under the stage," he mumbles.

"Jesus fucking Christ," she curses. "And you were on that stage! What *was* that anyway?"

Cooper is quiet, his eyes cast down. "You think they got the message?"

She shakes her head. "I think the bomb kinda stole your thunder. But it was a pretty solid speech."

On the opposite corner of campus, Pando and Iri reconvene at the 1 train. Her arms folded into a tight screw, she contemplates her subordinate with superlative indignation.

"Iri, I don't know what went wrong," he swears. "I pushed that button with all my might. Nothing happened. I'm serious."

She nods grimly. "Mine didn't work either. Let's get back to BKP. We'll figure it out there."

The two part ways, Iri heading to regroup with the other agents. Pando pulls out his phone, and sees an icon for one missed call. It's from Shree.

Fuck, that was not a good time to call, he thinks, upset that he hadn't the chance to answer. He dials her number, only for it to cut straight to voicemail. He hangs up in frustration.

Cooper: Natasha? Are you ok?? Please get back to me 4:14pm
Natasha: I'm ok. Are you? 5:33pm
Cooper: Yeah, I think so. 5:33pm
Natasha: You should check the news. Channel 4. 5:34pm

From the warmth and safety of his mattress on the floor, Cooper opens his computer and watches the live stream broadcast. The lower-third reads "TERROR AT COLUMBIA" as frightened students get interviewed on screen by News 4 reporters.

"It's insane how ruthless these anti-environmentalists have become," opines a young fair-haired woman. A wave of unrest has been kicked in motion by the day's events; BKP's strategy is proceeding as planned. The revolution is already in motion.

He still can't believe Pando didn't pull the trigger on him. His reckless gamble paid off, and because of him, nobody died today. Well, almost.

He sends Pando a text, frustrated that he still hasn't heard from him.

Cooper: I need to talk to you. 5:38pm

He turns back to the broadcast, watching as the hyperbole plays out on screen. His attention is broken by a buzz, though not the kind from his phone. It's the urgent sound of the doorbell.

He slips on a shirt and descends the gray braids of rope. He approaches the door skeptically, uncertain as to who could be knocking at this hour. He peers through the peephole; the distinct flash of a badge against navy blue is all he needs to see.

The officer asks to come in. Nervously, Cooper motions her inside, prepared to plead the Fifth if she asks any probing questions about his whereabouts in the past 24 hours.

"What's all this about?" he asks the cop.

"A body was found in the Hudson this afternoon," the officer explains. "A young woman, about your age. This address was listed on her ID card." The policewoman holds up a photograph of the card, featuring the name and picture of one Sharifa Morgan Dunlap.

Cooper's eyes nearly bulge out of their holes, as if he has witnessed the long, streaky flame of an asteroid overhead. A blush of nausea overwhelms his senses when he sees her smiling face on the card. He feels a subtle movement in his gut, like the quick legs of a spider crawling up his stomach lining. A putrid taste enters his mouth, and he begins retching into the concrete.

Curdled gobs of pink spew from his mouth into a swampy puddle by the officer's feet. He coughs out the rest, bringing an end to the attack. He squeezes his eyes tightly; a tear clings to his cheek as he wipes his hand over his dripping mouth. The marbled bile on the floor stares back at him derisively.

After a few minutes of the worst conversation of his life, Cooper sees the cop to the door. He throws himself onto the orange cushions in disbelief; this day has been so surreal that he can't be sure if he's dreaming or not. Too stupefied to function, he gazes up at the ceiling, his vision going out of focus. He lays there for a minute, waiting for the tears. It always seems that in moments like this, they refuse to come at all.

Nearer My God to Thee

Cooper awakes to a godless world. His body is as heavy as a gravestone, the bed beneath him his coffin. The casualty list stands at three. And that's just counting the humans.

Futility lies on the floor by his feet. He sees it there, insulting him. He stares at it with ire.

"Fuck you!" he shouts, leaping from his bed. He snatches it from the floor savagely and storms out of the apartment with it. He hoofs up the stairs to the roof, where he goes right to its edge and looks down at the street below. The cloudy sky watches the drama unfold as it rolls by listlessly above.

In a fit of primal emotion, Cooper hurtles the book across the sky with all the power he can summon. He watches as it falls, sure as gravity, onto the empty sidewalk below, liberating him from the accursed story.

Atlantis looks back at him against the ash gray sky, sadder today than ever he's ever seen it.

Despite his crippling sorrow, he allows himself to attend class on this last day of college. The people on campus are rightfully wound up, wary faces glancing in every direction. This alone is enough to set Cooper on edge, but when he starts getting visibly scrutinized by the occasional student passing by, he begins to wonder if they know something he doesn't want them to.

A girl walks up to him in the hall. "Oh my god," she exclaims, "you're the one from the protest yesterday."

Cooper gulps dryly. "Uh, yeah, that was me," he admits.

"I really liked what you said."

The compliment takes him by surprise. "Oh, thank you," he fumbles.

She beams at him earnestly. "It was really inspiring. You really opened my eyes, and I'll never forget it. For the first time in a long, long time, I feel hopeful."

He looks deep into her hazel eyes, profoundly moved by her heartfelt praise. "That means a lot to me," he clucks. Words can make an impact after all.

This ends up being no isolated incident. He gets stopped in the hall no less than six other times over the course of the day, his fellow students commending him for having the gall to say the things he did. The attention is overwhelming.

Natasha: Your breaking the internet right now Coop! 1:12pm
Cooper: Huh? What do you mean? 1:15pm
Natasha: Your speech from yesterday! Its going viral! 2 mil views and counting!! 1:17pm
Cooper: Really?? Do you want to meet up later? 1:18pm
Natasha: I can't, the club and I are already drafting up a new campaign, #greenlivesmatter. But I'll see you at graduation! 1:25pm

Cooper watches with casual indifference as the footage of his speech blows up in popularity, shared from one social media site to another as the digital generation responds to his captivating oratory. "*Columbia student steals stage and schools everyone with message of hope*" is the clunky title of his now-famous moment, recorded from a phone in the front row of the crowd. Whoever captured this video was positioned within the blast radius of the bomb.

In the days following his big moment, the video is viewed 14 million times, from every continent on Earth — even melting Antarctica. His words resonate throughout the blogosphere and into the realm of television. Each of the major television stations broadcasts snippets of his speech to the world, the final bumper before the commercials at the end of the nightly news. Not a day later, he's being requested for an interview.

"People are calling you the voice of a generation. How do you respond to that?"

Cooper dwells on it for a moment before responding to the reporter. "I wouldn't say that exactly, because there can never be just one voice to speak for a large and complex body like us. There have to be many voices."

"So you're like the conductor?"

He shrugs. "Sure, something like that. I could be the conductor, but I alone can't bring the performance to life. It's gotta be all of us. But I'll say this: if there was ever a generation that could save the world from climate change, there couldn't be a better contender than ours."

"If your movement succeeds, how would you want us to think of you then?"

"Hmm." He taps his chin with the ball of his forefinger. "I— consider me the albatross."

The interview airs nationally later that night. At last people are paying attention. At last we may still save ourselves. At last rationality may prevail.

But rationality wasn't enough to save Shree. And that's a wound from which he may never recover.

He feels a vibration from his pants. Hurriedly, he pulls out his slate. It has no notifications, no messages; just a phantom buzz, ringxiety getting the best of him.

Still no word from Pando.

. . .

Nevaeh sits at BKP, a stern look on her face. She's leering across the table silently, Iri reading off a paper by her side. Pando sits nervously opposite them both.

"And so this investigation concludes that Pando Lamar was in no way at fault in the incident in question, and he shall be restored to the full rank of Clean Front agent, effective immediately." Iri puts the paper down flat, revealing Pando's scarred face beyond.

He takes a deep breath. "You could have told me about the backup plan," he says. "It would have been nice to know."

Nevaeh shakes her head. "It was my decision," she says. "After Cooper turned, we couldn't take any chances."

"Always have a Plan B," adds Iri.

He nods ponderously. "So when do we tell the world about Coke?"

Nevaeh clasps her hands forward. "We're meeting with a reporter from the *Times* tomorrow morning to submit our anonymous tip. By then, every university in our territory will be protesting. And from there, it's war."

He grimaces approvingly. "Put me on the front lines," he says. "I'm ready, coach."

Iri flashes an agreeable grin. "We have bigger plans for you, Pando. More leadership. More important work. We're gonna need your skills in the days ahead."

He cocks his head, intrigued by this vague insinuation. "So what's the next move?" he asks.

The blonde woman pulls out a white sheet of paper with a map of Mexico inked into the fiber. "We found it," she reveals, placing her finger over the red X on the page.

Pando's eyes alight at the news.

"We tracked one of their vehicles to the new location," she explains. "It's just outside of Chicxulub, at the tip of the Yucatan. We're spinning up for a strike."

He stares intently at the red mark on the map, located at the very center of the ancient crater. "The asteroid strikes again," he muses. "It's like he knows exactly what he's about to do."

Iri nods fatally. "The Battle of Armageddon is about to begin."

Back in Bushwick, Jeremy Braydon is in town for his son's graduation. He sits on the couch in the living room while Cooper slides into his blue gown.

"Looks like showers in the forecast," says his dad, staring blankly at the tablet in his hand. "Chance of rain 60%. Better bring an umbrella."

"On graduation day?" asks Cooper, disinterested. "That bites."

Jeremy shrugs. "You can't control the weather."

Cooper pulls his arm through the heavy garment, his hands barely making it past the cuffs. Here he is at last, graduation day. He never thought he would make it.

He thinks back to the start of the year, how eager he was, how he just wanted it to be over already so he could start his life. But now look at him — he doesn't want it to end. He doesn't want to grow up after all.

"Ya know, I'm proud of you, son," Jeremy boasts, marveling the young gentleman in his robe. He rises from the couch to stand before his son, grown to his same height. "It feels like just yesterday I was teaching you how to ride a bike, and now here you are, a grown man. You even got a few hairs on your face."

The words imbue Cooper with a sense of relief. If his dad thinks he's a man, then maybe he's ready for the real world after all.

"Thanks, dad," he says.

"Listen, Coop," Jeremy goes on. "I've been thinking about it for a while, and I've decided that if you want to go to grad school for sustainability, then you can go. It's your life, and you should live it the way you want. You're old enough now."

Cooper's eyes brighten, touched by his dad's generosity. There could be no better thing for him to hear in this moment. His face cracks into a smile.

"But what if I don't make it?" he asks after a moment, considering the possibility of failure. "Will you take me back in if I need you?"

Jeremy laughs a deep, hearty laugh. "You won't need me, Coop."

This indefinite answer disappoints Cooper. His eyes turn to the floor.

"But yes, if you ever need anything, I'm here for you," Jeremy adds. "You're my son."

Cooper looks up again, and for the first time that he can remember, he and his father share a long embrace, the oxytocin in their brains binding them together like glue.

As they head out the door, Cooper checks his phone; no new messages. Who knows if Pando is still even alive? He decides to text him anyway.

> Cooper: Hey man, I hope you're ok. Just wanted to let you know, Shree's funeral is Monday. She would want you to be there. 7:29am

He stares back at the words on his screen, pixelated virtuality turned all too real. He puts aside the discomfort of speaking for a dead person and follows his dad down the stairs.

The skies on this graduation morning are dark and gray, an inescapable pall drawn over the world like a veil. Jeremy sits in the rows of plastic chairs with the rest of the crowd, watching from afar as he hears his son's name come over the speakers.

"Cooper Braydon," the dean announces.

The young man walks up to the stage. A cloaked woman hands him a degree that rests empty in his hands, a hollow symbol to prove to the working world that he's earned something. As he steps off the stage, he sees storm clouds gathered on the horizon.

When the ceremony concludes, Natasha finds him in the crowd. "Hey!" she exclaims as they enter a tight embrace. "Congratulations!"

"Thanks," he says, looking down forlornly at his degree. "I guess I earned it."

She rubs his shoulder affectionately. "You're so smart now!" she proclaims.

He chuckles somberly. "I don't feel so smart," he says, looking into her eyes. "Do you wanna grab coffee tomorrow?"

She smiles at him warmly. "Sure. My place?"

His cheeks go hot; he's never been invited over before. "Yeah, that'd be great," he agrees.

"I'll text you."

He turns away, searching the crowd for his dad. As he looks around with postgraduate eyes, he realizes he now faces the most daunting challenge of them all: making himself relevant. For the first time ever, he is staring down the infinity of his life. There are no more structured years ahead of him, no new classes in the fall, no predetermined path for him to follow. He now walks the endless stretch of days that lie before him, each one unnumbered, pulling him toward an uncertain future of unknown possibilities. A world where his degree is both essential yet insignificant. He's now just like everyone else: an undifferentiated college-educated white guy that can do nothing and knows nothinger.

Still, with $200,000 degree in hand, he feels he has finally moved beyond adolescence. He's reached the end of the beginning. Now it's time for the beginning of the end.

. . .

Steaming coffee sloshes playfully into a white mug, its sound music to Natasha's ears. She pours another cup for her friend, bringing both to the dining table in the middle of her kitchen.

"Thanks," says Cooper, accepting a mug. He stirs it vacantly while she sips her hot drink.

"So are you gonna keep working at Straunt now that you're a celebrity?" she asks.

He shrugs lazily. "Probably," he says. She sips her drink, waiting for him to elaborate, but he never does. She thinks of something else to fill the silence.

"Did you hear that Columbia's drafting up a plan to divest?"

He nods as he takes a sip, scorching liquid passing down his throat. "Small victories," he grumbles.

After a moment of awkward silence, he finally decides to speak his mind.

"Listen, I have to tell you why I did what I did on that stage," he says. He can't take credit for saving her life if she doesn't know that he saved it. "But you have to promise not to tell anyone."

Natasha stares back at him as she sips from the mug. She says nothing.

"I knew there was a bomb," he confesses.

She smiles calmly, the exact opposite of her anticipated reaction. "I know you did," she breathes.

Cooper looks at her like she just admitted to being Jesus. "What do you mean?" he asks. How the hell could she know that?

"Because I knew too."

"What?" he cries. And then it dawns on him — it's in her eyes, and maybe even in her smile. The more he thinks about it, the more it makes sense. And though he never put two and two together, with graver issues occupying his thoughts, he now realizes the obvious truth: Natasha Brooks is one of the Four Sisters of BKP.

"You gotta be shitting me," he says.

She smiles knowingly at him.

"You were part of this the whole time?"

"Of course."

Through his eyes, she has morphed into some hideous monster, an imposter removing her human skin like a glove, revealing the ugly beast inside. How could a girl as sweet as her give in to murder? It doesn't make any sense. Then again, neither does burning carbon.

"You were gonna be a victim of your own cause?" he asks in disbelief.

She sits back in her seat. "We're all *victims*, Cooper, all of us. And we're completely powerless unless we start taking some risks. Some *real* risks. Life or death risks. There's too much at stake, and no time left."

"Don't patronize me, I know where we are," he snarls.

"Then you also know what we must do," she begs. "And you can still be one of us."

He shakes his head skeptically. This is unbelievable.

"You're still gonna go through with the war?" he asks hopelessly. "Even after everything I said?"

She looks at him like he's a fool. "Of course we are," she says. "You may have sabotaged our bomb, but you didn't stop the boom. The public would've lost interest in this scare days ago if your video hadn't gone viral."

A flash of pride creeps onto his face. 40 million views and counting.

"But because you're 'The Albatross' now, people are paying attention, just the way we planned."

So he pulled it off after all. A revolution of words, not blood.

"You're the only one who defied our expectations, Coop," she tells him. "The only one whose numbers added up, yet didn't adhere to the plan. You're an anomaly. And we need you."

The only one, he thinks. He isn't some mere automaton, quantified by the raw data of his life; he's a human being, with all the imperfections it entails. And he's decided once and for all that it's going to mean something more than just being nature's perfect murderer. With this compliment under his belt, he feels like a new ship, capable of any heading. The power of free will circulates through his blood, the wisdom in his mind to know what to do with it.

"You're one of the bravest people I've ever known," she fawns, "and the Brooklyn Project needs people like you, who care as much as you. You're one of us."

He stares into the soon-to-be extinct coffee in his mug. "If you people want to die, then so be it," he responds, looking up. "But me? I want to live."

Natasha's face turns melancholy as it reflects off the brew in her cup. "It's true, I was ready to die," she admits. "But if that's what it takes, then it's what I'm prepared to do. I'm not going to let them ruin everything I love, everything my mother loved. I'm just not." Her face crumples, sadness swelling within her. A tear tumbles down her caramel cheek.

He continues to shake his head, still in disbelief. She's just like him. Then again, so is everyone.

"I'm not one of you," he decides. "And I'll never work for BKP again."

She really thought she could sway him to return. But seeing him here, stubbornly turning away from his destiny, she realizes she was wrong. Her eyes go dull, her heart deflating. The game is over.

"But I won't do anything to stop you either," he adds.

She looks at him quizzically. "That's not gonna fly," she says. As long as he knows what he knows, he's too dangerous to be left to his own devices.

Cooper leans forward in his chair. "It's gonna have to," he threatens, "or I'll start spilling the beans before you can even figure out where I am."

Her eyes narrow distrustfully. "Then you have to promise me — *promise* me — that you say *nothing*. To anyone."

Careful not to break eye contact, he pledges his word. "I promise."

She sighs unhappily as she shakes her head with contempt. "You really can't stop us, Cooper," she says. "All you can hope to do now is to beat us to the punch."

He rises from his seat, his eyes still trained on her. "I already have," he breathes.

She frowns in disagreement. "The climate wars have begun, whether you believe it or not." She glares at him with immutable conviction.

He squints into her cocoa eyes. They look every bit as bright and intelligent as when he first saw them. But they have lost all meaning to him; now she's just another human.

He shakes his head slowly, resolutely. He bends down over the table, getting right up in her face: "No."

Straightening for the exit, he turns the handle and walks out, the door closing loudly behind him. He never wants to see her again.

. . .

The ceremony on this Monday afternoon is quiet and short. A handful of Shree's friends and family are in attendance, her weeping mother the most conspicuous of all. Though they've gathered in a cemetery, there is no body to inter; as outlined in her suicide note, Shree's cremated remains have been packaged into an eco-urn. She selected the seed of a pine for her ashes to nourish, so there can be green on her tree all year long.

Mel sniffles alongside Cooper after the conclusion of the ceremony. He wraps his arm around her consolingly.

"I'm so mad at her," she says through tears. "Why did she do this."

Cooper rubs her arm sympathetically. In the span of a month, she's lost two of her closest friends. "I forgive her," he says. "I just wish I had been there for her."

Mel pulls out a handkerchief to blow her nose. "I— I gotta go to the restroom," she says, about to lose her composure. She kisses him on the cheek before turning away. Cooper watches as she leaves in solitude.

As the rest of the mourners begin to clear out, Cooper recognizes a face in the back of the group. It's Pando; he showed up after all. With a pinch of anxiety, he walks over to his friend for the first time since the protest.

"Hey," Cooper says on approach.

"Hey," the refugee offers in return, sadness in his heart.

"Everything alright?"

Pando nods.

Cooper nods back, relieved to see his brother again after so long. But in light of the circumstances, it really isn't enough.

"I can't believe she's gone," Cooper muses, his foot kicking the grass.

"Me neither, man."

Cooper tries to hold back the tears, shaking his head dismissively as he fights them off.

"How many people are going to die?" he asks the clouds.

Pando studies him, thinking for a moment about the question. "How many people are alive?" he asks.

"Billions."

He nods. "Alright then. Billions."

Cooper sulks. "I can't believe it actually came to this," he croaks.

Pando looks over to the freshly-planted patch of dirt, exhaling a long sigh. "Shree killed herself because she didn't think there was anything left to do. She thought she was just another human stomping over the Earth."

"But she *wasn't* just another human," Cooper cries. "She could've been so much more."

Pando's eyes lose focus. "We're all killing ourselves, Coop, all of us," he thinks aloud. "She just got it over with early." Then, his head held low, he succumbs to despair.

"This is all my fault," he says. "She wouldn't even know about climate change if it wasn't for me."

"That's not true, Pando."

"It is true," he holds. "I'm the one who sent her down this road. It was me."

Bleariness fogs his vision, the stone dam holding back the tears beginning to buckle under the weight of the emotion behind.

"I wasn't there for her when she needed me," he admits. "She called me during the protest, but I was busy trying to kill you."

Cooper is nonplussed. "I know you were," he says.

Pando reacts mildly, happy to see that after six months, his friend knows him well enough to know when he'd be willing to murder him. And that's when it hits him.

"God, I'm so sorry, Cooper," he wails. "I'm sorry for what I had to do. But fuck, man, I'm not gonna let 'em get away with it. I'm not gonna let 'em win." His stoic face finally cracks, an unstoppable flood of human feeling draining down his cheeks.

"I'm gonna keep it in the ground," Pando avows. "I'll put my body in front of their machines if I have to. I'll do anything. Even if it means killing my brother. Because I've got to stop them. For Shree. For everyone. For everyone who's ever lived and everyone who ever will. I won't let them burn it all. I won't."

The cold splash of a raindrop lands on Cooper's forehead. He looks up into the bone gray sky, its dark clouds pregnant with moisture. For the first time in months, it begins to rain.

"I forgive you," Cooper sniffs. "This is bigger than me and I know that." He again looks up at the mottled sky. Somewhere above the clouds, his mother is looking down on him. So is Shree.

The American flag in the distance catches the breeze from halfway down the pole. It's Memorial Day; Cooper can't help but think not just of our fallen soldiers, but of our soldiers yet to fall. The ones who will die in the wars of tomorrow, fighting the mistakes of today. The world we leave them will be more treacherous than at any point in history. He realizes Natasha's right: the climate wars have already begun.

Yet as an El Niño brews quietly in the waters of the Pacific, and as Earth passes through its warmest May on record, Cooper feels a pang of hope. Maybe Shree's suicide wasn't for nothing. In her embrace of death, she has reminded him just how important it is to cling to life. And as long as there's still a world worth saving, his life has purpose, his existence has meaning. It isn't over till it's over.

He's learned how to die. Now he's ready to learn how to live.

"So what do we do now?" asks Cooper.

Silver eyes glint at him. "You know what I'm gonna do," Pando says. "I'm gonna fight. There's still a geobomb out there. The Brooklyn Project has big plans ahead, and I'm gonna be part of them. We're gonna give the human race a chance."

Cooper nods in grave understanding.

"And you?" asks Pando.

"Same as you," he replies. "I'm gonna keep it in the ground. And I'm gonna save the world before you do."

Pando smiles at his friend. "Two degrees, right?"

"Two degrees."

The rain begins to fall harder now. It feels good on Pando's skin. Water. Cleansing, life-bringing water. He looks to the tiny sprout in the soil. Grow, little tree.

Cooper thinks back to earlier, happier memories. The two men stand silently, sharing a moment in the sprinkling rain, reflecting on all they've been through, and all they're about to go through. The real fight is only just beginning.

"So what are we waiting for?" he asks finally.

Pando shakes his head, determination in his heart. "Nothing," he breathes. "It's on." He holds out his hand; they shake firmly.

"May the best Man win," Cooper says. "Let's get to work."

So they got to work.

ABOUT THE AUTHOR

J.R. Kirsch is a climate activist and court reporter.
He thinks it's time to talk about climate change.
He hopes this book will help facilitate the conversation.

CPSIA information can be obtained
at www.ICGtesting.com
Printed in the USA
FFOW02n2255230816
27095FF

9 780997 469301